Agribusiness:
An
Entrepreneurial Approach

Agribusiness: An Entrepreneurial Approach

by

William H. Hamilton, Professor Emeritus
Agricultural Education, Purdue University
B.S. and M.S., Iowa State University
Ph.D., Cornell University

Donald F. Connelly, Vocational Agriculture
Instructor
Western High School, Russiaville, Indiana
B.S. and postgraduate work, Purdue University

D. Howard Doster, Associate Professor
Agricultural Economics, Purdue University
B.S., M.S., and Ph.D., The Ohio State University

 Delmar Publishers Inc.®

NOTICE TO THE READER

Cover Design: Mary Beth Vought
Cover Illustration: Jeanne A. Benas

Delmar Staff

Senior Administrative Editor: Joan Gill
Project Editor: Andrea Edwards Myers
Production Supervisor: Karen Seebald
Production Coordinator: Sandra Woods

Art Manager: John Lent
Art Coordinator: Michael Nelson
Design Supervisor: Susan Mathews

For more information address Delmar Publishers Inc.,
2 Computer Drive West, Box 15-015
Albany, New York 12212

Copyright © 1992
By Delmar Publishers

Printed in the United States of America
Published simultaneously in Canada
by Nelson Canada,
A division of The Thomson Corporation

10 9 8 7 6 5 4 3 2 1

Library of Congress Cataloging-in-Publication Data

Hamilton, William Henry, 1919–
    Agribusiness, an entrepreneurial approach / by William H.
Hamilton, Donald F. Connelly, D. Howard Doster.
        p.   cm.
    Includes index.
    ISBN 0-8273-4447-3
    1. Agricultural industries—Management.   2. New business
enterprises—Management.   3. Entrepreneurship.
I. Connelly, Donald F.   II. Doster, D. Howard.   III. Title.
HD9000.5.H35 1992
630'.68—dc20   90-47237
                                                                          CIP

CONTENTS

PREFACE

ACKNOWLEDGEMENTS

PART I.   A FIRST LOOK AT ENTREPRENEURSHIP

1   SHOULD I WORK FOR ME OR FOR SOMEONE ELSE?   1
     What Factors Should I Consider In Deciding To Work
     For Myself Or For Someone Else?   3

PART II.   PLANNING FOR ENTREPRENEURSHIP

2   HOW DO I START A BUSINESS PLAN?   15
     A Business Plan Serves As A Road Map   19

3   HOW DO I SET PERSONAL AND BUSINESS GOALS?   27
     Goal Setting   30

4   WHAT TYPE OF BUSINESS ORGANIZATION SHOULD I CHOOSE?   41
     Types of Business Ownership   43

5   SHOULD I PURCHASE A BUSINESS OR BUILD ONE?   51
     Starting in Business   53

6   WHAT ARE MY THREE BUSINESS RESOURCES?   59
     Identifying Your Resources   61

7   WHAT IS MY COMPETITION?   69
     What Causes Competition   71

8   HOW DO I DEVELOP MY MARKET NICHE?   79
     What Is Market Research?   82
     Should I Research My Market?   83
     How Do I Research My Market?   83
     Plain or Fancy Research   86
     How Do I Expand My Product or Service?   86
     How Do I Develop New Accounts?   88
     How Do I Develop New Products?   89
     You've Developed Your Niche   89

9   HOW DO I PROMOTE MY BUSINESS?   93
     Promoting Your Business   95

PART III.   ACCOUNTING FOR ENTREPRENEURSHIP

10   HOW DO I ACCOUNT FOR BUSINESS TRANSACTIONS
     AND EVENTS?   103
     Accounting For Your Actions   105

11   HOW DO I USE A SINGLE ENTRY SYSTEM?   117
     Making Records Simple   118
     Record Preservation   121

12   HOW DO I USE THE DOUBLE ENTRY SYSTEM?   125
     Entities   126

13   DECISION, DECISIONS...HOW DO I MAKE THEM
     DURING THE START-UP PERIOD?   143
     Detailed Cash Flow Analysis   148

14   IN DECISION MAKING, HOW DO RECORDS HELP ME
     PROJECT CASH FLOW?   155
     Records and Reports in Decision Making   157

15   HOW CAN I MANAGE CUSTOMER CREDIT AND COLLECTIONS?   165
     Managing Credit   168

PART IV.   MANAGING ENTREPRENEURSHIP

16   HOW DO I MANAGE THE LEGAL ASPECTS
     OF A BUSINESS VENTURE?   175
     The Law And Your Business   176
     The Need For A Company Lawyer   181

17   HOW CAN I MANAGE TAX RESPONSIBILITIES
     FOR THE BUSINESS VENTURE?   183
     Looking at Taxes   184
     Are Taxes Fair?   185

18   WHERE SHOULD MY BUSINESS BE LOCATED?   189
     Looking At Locations   190

19   WHAT FACILITIES DO I NEED FOR THE BUSINESS?   197
     Business Facilities   199

20   WHAT ARE THE COSTS OF RUNNING A BUSINESS?   203
     Business Costs   206

21   HOW DO I PRICE MY PRODUCT OR SERVICE FAIRLY?   217
     What Factors Influence My Pricing?   218

22   HOW DO I SCHEDULE MY PRODUCTION, SERVICE,
      OR SALES ACTIVITIES?   231
         Cash Flow   234
         Credit Regulations   237
         Scheduling Production   237
         Solving Problems   239

23   HOW CAN I ESTABLISH A SOUND SET
      OF BUSINESS PROCEDURES?   241
         Establishing Business Standards   243
         Delegating Work and Responsibility   252

24   HOW DO I DEVELOP CUSTOMERS?   257
         Developing Customers   259

25   HOW DO I MANAGE PERSONAL CONTACTS?   265
         Managing Business Contacts   266

26   WHAT SHOULD I CONSIDER WHEN HIRING OTHERS?   271
         Steps In The Hiring Process   274

27   HOW DO I MANAGE BUSINESS RISK?   283
         Self-Quiz—What Is My Risk Rating?   287
         Managing Risks   288

28   IS THERE A BASIS FOR PROFITABLE COMPUTER USAGE
      IN MY ENTERPRISE?   299
         Business and Computers   302

29   HOW DO I MANAGE BUSINESS GROWTH?   307
         When Should You Expand?   309

30   SHOULD I CHANGE THE BUSINESS STRUCTURE?   315
         Change In A Business's Structure   316

31   SHOULD I SELL THE BUSINESS?   321
         Selling The Business   323

PART V.   ANOTHER LOOK AT ENTREPRENEURSHIP

32   A LAST LOOK AT ENTREPRENEURSHIP   329
         The Personal Satisfaction in Self-Employment   330
         Writing a Business Plan   331
         Goal Setting   331
         Organizing Our Business   333
         Obtaining A Business   334
         Market Research   335
         Decision Making   337

	Advertising	337
	Competition	337
	Accounting Procedures	338
	Rules, Regulations, and Taxes	338
	Costs and Prices	338
	Scheduling Business Operations	338
	Business Procedures	339
	Human Relations	339
	Business Risk	340
	Computer Use	341
	Changes in the Business	341
	Is Entrepreneurship For Me?	341
33	AM I SUITED FOR ENTREPRENEURSHIP?	343
	What Skills Do Entrepreneurs Need?	347
	How Can I Assess My Potential as an Entrepreneur?	348
	GLOSSARY	351

PREFACE

Entrepreneurship starts with me—I have an idea! Many people in the past have had that same uplifting experience. We can learn from them and from their experiences. That's what this book is all about. Your idea could be a big winner and could provide your living and more. You must start where you are today with your resources. You set your goals to obtain what you want most.

One objective in life is to manage yourself to get the most possible satisfaction from the use of your resources. Right now you have three resources: mind, muscle, and money. Most of you probably have more muscle and mind than money to use right now in your efforts to reach your goals. Some people do a better job of managing their resources than others; they get more satisfaction from the use of their mind, muscle, and money. We say these people are better managers.

Do you want to be a better manager? Of course you do! No one wants to waste anything. Even when we choose to relax, we want to get the most fun from that time.

How can you improve your management skills? By taking this course, you will:

- learn more about yourself.
- learn more about your resources and how to use them.
- learn how to make choices.
- learn business management skills.
- learn how to create a business plan.
- complete interesting activities that will help you understand our American economic system.

We begin this study with a look at business types, their forms of organization, and the advantages of entrepreneurship. We progress through what is needed to start a business enterprise, and how to understand business accounting. The text moves through the details of business management and then completes with a summary and a self-evaluation for entrepreneurship.

An entrepreneur is a person with a burning idea. This person won't be satisfied until the idea is developed. Try out your great idea and become an entrepreneur. If you decide not to test your idea, this text will give you a better understanding of how businesses operate in the free market economy.

The Authors

ACKNOWLEDGMENTS

The authors express grateful appreciation to many people who have contributed to the completion of this text.

We would especially like to acknowledge the valuable contribution of our field test students who helped so much in the polishing of our work. These students were from Western High School, Russiaville, Indiana. The students involved in the pilot test were:

Matt Arbuckle
Shane Burnette
Sharon Burton
Jason Dwyer
Greg Dyer
Kenny Elmore
Brett Etherington
Tracy Etherington
Reg Gilbert
Vance Hale
Larry Houston
Manson King
Rick Lawson
Danny McCain
Ronnie Orem
Karyl Pogue
Lance Rivers
Bruce Rood
Joe Russeau
Tony Settle
Brian Sewell
Mike Sherrod
Jason Smith
Beth Spoon
Jeff Tillman
Tim Young

The authors would also like to acknowledge with thanks the review of the text by:

Joe Waldman, Jr.
Deweyville High School
Deweyville, Texas 77614

Dr. Richard Churchill
Southern Maine Technical College
South Portland, Maine 04106

David Hall
Tucumcari High School
Tucumcari, New Mexico 88401

Gerald Mc Donald
Spring Branch Education Center
Houston, Texas 77055

Donald Connelly expresses appreciation to Patricia L. Brettnacher for her excellent writing skills in developing Chapters 2, 3, 8, 9, 20, 21, 22, 23, 26, and 28. He also expresses appreciation to his wife Janet for her support throughout the project.

Howard Doster expresses appreciation to his secretary Betty Cottrell for typing the manuscript and his wife Barbara for her support during the writing of the manuscript.

William Hamilton expresses appreciation to Donna Barratt for her many hours in typing, editing, and preparing the final copy of the text. He also expresses his appreciation to his wife Barbara for her support during the project.

Donald F. Connelly
D. Howard Doster
William H. Hamilton

CHAPTER 1

Should I Work for Me or for Someone Else?

OBJECTIVE

To identify key concepts of entrepreneurship and the various forms it takes.

COMPETENCIES TO BE DEVELOPED

After completing this chapter you will be able to:
1. Define the terms to know.
2. Identify at least 10 characteristics of successful entrepreneurs.
3. List a dozen advantages of self-employment.
4. List the advantages of working as an employee.
5. Name five reasons that businesses fail.
6. Identify four forms of entrepreneurship.

TERMS TO KNOW

Adversity
Black cloud
Case study
Cash flow
Commitment
Distraction
Economic downturn
Entrepreneur
Equity
Funded

Profit
Purveyors of doom and gloom
Regress
Resource
Risk
Supervised Occupational Experience Program (SOEP)
Venture
Viable

INTRODUCTION

In this introductory chapter, we will explore some ideas about your place in the world of work and identify the *resources* of a business venture. In the process, you will find some words that are new to you and some whose meaning isn't clear. At the end of book you will find a glossary containing simple explanations of words like those listed under terms to know.

CASE STUDIES

The following real-life *case studies* illustrate several important principles and economic concepts that we will explore as we discuss starting a small business enterprise. Remember these people—you will meet some of them again in later chapters.

The Case of the Sweet Deal

Kermit, as a high school sophomore, owned a sow and litter that he kept at his uncle's farm where he worked most of the time. His mother died at his birth, and his father was killed in Vietnam, so he was raised by his grandmother and uncles.

Kermit decided to add sweet corn to his Supervised Occupational Experience Program (SOEP) and budgeted a potential *profit*. With the aid of his grandmother and agricultural education teacher, he rented three acres of ground not yet sold by the developer of his grandmother's subdivision. He traded labor for use of his uncle's machinery, and the project was a great success. He was persistent in his sales work, and he realized that he could always feed any surplus to his pigs.

Kermit easily won the State Farmer recognition as a senior.

The Case of a Black Cloud With a Silver Lining

Bill Brent completed his degree in agricultural education, and taught it for two years. He was a failure at discipline in the classroom, so he quit and began farming with his father who wanted to retire in a few years. His father sold Bill the machinery that, as on many other farms, was over abundant. In three years, the banker predicted an *economic downturn* coming and refused to advance the operating capital Bill needed to continue. The banker told Bill his *cash flow* wouldn't carry his machinery debt load, his operating costs, and his family living expenses. The land was all rented, mostly by cash rent. As a result, Bill was forced to sell out; fortunately, when the sale was completed, he paid off all debts and had a sizable sum to invest. He has been working in a mid-level job in the nearby county seat while he looks for the right opportunity to strike out on his own. He thinks his *black cloud* turned out to have a silver lining. Had he managed to continue, in two years his *equity* would not have paid his debts and he would have been wiped out in the economic downturn. He now feels the banker was a friend, not an enemy as he thought at the time.

Examining the Cases

What lessons do we learn from the experiences of these people? Some questions to think about as we read are:

1. What are the characteristics of successful *entrepreneurs?*
2. What factors should we consider before starting a business *venture* of our own?
3. What advice would you offer Bill?
4. What abilities or skills does an entrepreneur need to succeed?

From the case studies, these facts seem to be important for consideration.

1. Kermit was forced to mature early, and had the self-confidence to tackle a moderate *risk* venture.
2. Kermit did not rush blindly into the enterprise—he planned well first.
3. Kermit used the expertise of others.
4. Kermit made contingency plans to lessen the risk (use corn for pig feed).
5. Kermit made a *commitment.*
6. Kermit was persistent—he planned his work and worked his plan.
7. Kermit bartered (traded) work for equipment to keep his costs down.
8. Bill evaluated his performance, and took action that made sense to him.
9. Bill learned that bankers like plenty of collateral for loans—the easy loan to get is the loan that is hardly needed. Bankers are not permitted to be heavy risk takers. The money they manage belongs to the public (their depositors) and therefore, their loans are subject to bank examiner review.
10. Bill is, like most of us, slow to take decisive action. He is still searching for the right opportunity.

WHAT FACTORS SHOULD I CONSIDER IN DECIDING TO WORK FOR MYSELF OR FOR SOMEONE ELSE?

We humans are usually motivated by our needs, preferences, and desires. The style of life we want to live, our abilities, training, experience, and our financial circumstances lead us to set goals in order to satisfy our internal drives. Clint knows that if his lawn business is to prosper he must be ready, Figure 1–1.

The importance of our need for security, self-expression, recognition from others, and affection from others, all affect our decisions relating to self-employment or working for others.

Only one of the two people in the case histories felt that the risks were in the proper range, and an entrepreneur emerged. The second lacked the element of self-confidence that says: "Damn the torpedos, full speed ahead." Does this mean that if we lack the wish to create an enterprise of our own, that we are somehow forever flawed? By no means. If everybody went into business, we probably would *regress* to the point of producing our own basic food needs.

Figure 1–1   Clint gets an early spring start on tune-up and maintenance for his lawn business. (Courtesy Donald F. Connelly)

It has only been since agriculture provided food above the family needs, that barter made possible industrialization and today's high standard of living. In the early 1700s, four out of five workers were entrepreneurs. This percentage dropped to one in five by the late 1940s. From the 1950s to the 1980s, the percentage of people who are self-employed is again increasing.

Figure 1–2 represents the function of resources in enabling the entrepreneur to conduct a business venture. Most of you will find a way to substitute time (muscle) for money in a small business venture, but it is not always possible to do so. Mind represents the idea or spark that sets off the enterprise in the first place. It also represents the management skills needed to keep the business functioning, and the planning needed to obtain the financial backing to produce the product, and to keep the business going until sales are high enough to sustain the business activity and return borrowed capital. Money represents the input needed to purchase raw materials, seed, feed, fertilizer, and fuel to produce the livestock, crops, or manufactured products. Money also represents the costs of supplying the service on the part of the business. Muscle represents the actual labor to produce the service or product sold.

For example, the cost of the lawn mower, gasoline, and telephone calls for a lawn mowing service would represent the money portion of the triangle. The mind portion would be the contacts to line up the lawn mowing jobs, while the muscle would be represented by the actual mowing activities.

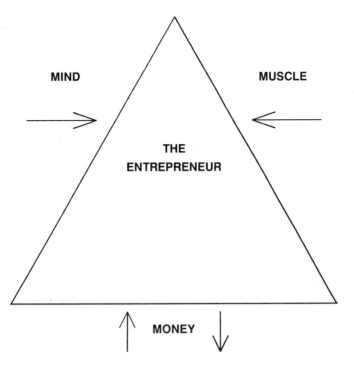

Figure 1-2 The Three Entrepreneurial Resources

Successful entrepreneurs often are those who are strong in these traits:

- *Problem Solvers*—Successful entrepreneurs like the challenge of problem solving. They welcome opportunities to exercise their creativity, energies, and expertise.
- *Low Need for Status*—They are not overly concerned by the opinion and regard of others. They know they are right, and that carries them through.
- *High Energy*—Entrepreneurs have good health and high energy levels. They are seldom sick—they just don't have time to be sick!
- *A Sense of Urgency*—They are restless unless they are working and may often seem to want challenges when everything is going too smoothly.
- *Self-Confident*—They know their own abilities and believe in their ideas.
- *Organization*—They organize their work, not always as others think it should be, but in their own logical system.
- *Vision*—Successful entrepreneurs can keep track of the whole enterprise and visualize how diverse parts fit into the whole operating scheme. They have command of the whole range of the business.
- *Flexibility*—If one option falls through, they will find another way to do the job or finance the change.

- *Persistence*—The successful entrepreneur doesn't quit when the going gets tough. The 5 o'clock whistle may only mean that work can now be done without other *distractions*.
- *Optimism*—Entrepreneurs are natural optimists. Things are always going to be great.
- *Independence*—Successful entrepreneurs think they can do the job better than anyone else. They like to be in charge and control their own destiny.
- *Emotionally Stable*—They are not given to emotional highs and lows as moods change.
- *Commitment*—Entrepreneurs do what is necessary to insure the success of their ideas. They won't be quite as discouraged by the purveyors of *doom and gloom*.
- *Risk Taking*—Entrepreneurs are not daredevils, but they take reasonable risks. They are willing to risk their own abilities. They like pitting their abilities against *adversity*. Figure 1–3 gives the message, "Don't be afraid to fail."

Don't Be Afraid To Fail

You've failed
many times,
although you may not
remember.
You fell down
the first time
you tried to walk.
You almost drowned
the first time
you tried to
swim, didn't you?
Did you hit the
ball the first time
you swung a bat?
Heavy hitters,
the ones who hit the
most home runs,
also strike
out a lot.
R. H. Macy
failed seven
times before his
store in New York
caught on.
English novelist
John Creasey got
753 rejection slips
before he published
564 books.
Babe Ruth struck out
1,330 times,
but he also hit
714 home runs.
Don't worry about
failure.
Worry about the
chances you miss
when you don't
even try.

Figure 1–3 Don't Be Afraid To Fail. (Courtesy of United Technologies.)

What are the Types of Business Organization that I Might Consider Appropriate?

Sole Proprietorship—The entrepreneur does everything—all tasks from design to production to sales. He or she does all the work and pockets the profits or suffers the loss. Examples are a seamstress with a sewing machine, a boy with a snow shovel after a six-inch snow, or a girl raising and selling popcorn. The business enterprise may be full- or part-time. Profit is not the only motivation for an enterprise, but is often the most important. The ability to demonstrate that one's ideas are *viable* (or workable) is motivating, but, in most cases, the business is likely to fail without profit.

Partnership—Two or more individuals share the cost, labor, and returns from an enterprise. One or more partners may work in the business, or one may operate the enterprise while others are silent partners. These differences will be described in more detail in a later chapter.

Inventors—People who create a new idea, design and perfect a new product or process, and create a company to produce and market the item.

Duplicators—Those who operate franchises or look-alike businesses. They duplicate the product or service another has made into a widely accepted and profitable business venture. Examples include copy-cat outlets of McDonald's and Dairy Queen in the fast-food area.

Developers—Businesspeople who buy a business that is underdeveloped or underfunded (without the opportunity to grow) with the purpose of improving the business for resale at a profit.

Scalpers—People who buy products or businesses strictly as a speculative opportunity.

Company Innovators—Persons working for larger companies who generate a special project within the company as contrasted with taking the idea outside the company on their own.

Corporations—Investor-owned businesses that operate much as an individual but with government supervision.

Cooperatives—Businesses owned by producers or consumers to provide a service in their own interest and to maximize returns or to minimize costs.

What Do Businesses Provide the Public?

Businesses can be divided into several classes based on their contribution. These are described in brief as:

Production—An enterprise that produces or manufactures a product. Examples are a corn farmer or a tire manufacturer.

Service—An enterprise that performs a service for others. Included are such services as tree trimming, custom baling, fast-food stores, legal advice, etc.

Sales—Listed as a separate division, they might be classed as services by some people. They include individuals or companies who market produce, manufactured goods, or services produced by other enterprises. Sales opportunities include wholesale selling to retailers, retailers selling to the public, or individuals selling directly to consumers.

What Are the Chances of Survival of a Business Venture?

The importance of advance planning in starting up a new business venture cannot be overstated when the chances for a business reaching its fourth birthday are only about 50–50. The chances are that about 40 percent fail in the wholesale business or in manufacturing, and over 50 percent fail in retail trade during the first four-year period.

As you see from these facts, the small business pool is always stirring with people entering and exiting in large numbers. New businesses replace the old. In short, the longer a new business survives, the better its chances of survival and maturing.

Why Do Businesses Fail?

Business failures are frequent and numerous. Their reasons for failure are many. As one author points out, failure in business is no more likely than failure in any other human endeavor. Cited were the number of college entrants that do not graduate and marriages that end in divorce. Figure 1–4 shows a business that failed. Here are a number of the most common causes of business failure:

Sales—Failure to set or reach sales goals may be an important failure point. Unless we let people know we have the best and get them to buy, we will fail.

Management—About half of all business failures are attributed to owner incompetence or mismanagement.

Inadequate Funding—Many businesses fail to reach a profitable level in the first two years. Few small businesses are adequately *funded* to survive until the profitability stage.

Economic Cycle—Timing is sometimes bad for competition to enter the market with a new or different product. Often cited are the Chrysler Airflow and the Ford Edsel. The Minneapolis Moline ''new style'' tractors introduced after the ''U'' models is an agricultural example.

Bad Debt Losses—This can be another stopper in the business world.

Competition—Well-funded, efficient businesses may be able to produce items and sell them for less, or even sell at a loss for a time to eliminate competition. Foreign competition with lower production costs is now disrupting the American economy.

What Are the Advantages of Working for Myself?

- If I work for myself, I am the boss. Independence may be the major reason for small business ventures today.
- What I earn above expenses is all mine.
- My own decisions are the ones that affect my life.
- I can test my ideas in a real life situation.
- I can set my own working hours and schedules.
- Although I don't pay union dues, I probably will join several associations appropriate to my business.
- I will have a variety of tasks from design to production to sales.

I will not have to watch out for office politics.
- I set my prices, especially in a service business. This is not true for some businesses such as independent insurance agents.
- I determine production schedules and control inventory.
- I determine the product or service to promote.
- I control the quality and my reputation.
- I solve the problems.
- I hire, train, and fire any employees.
- I control the advertising and business image.
- I set all company policy.
- It is easy to start a small business.
- It is easy to quit a small business.

Figure 1–4 Businesses fail for many reasons. Finances and location are two common causes. (Courtesy The Three Entrepreneurs)

What Are the Advantages of Working for Others?

In a small business:
- I will know everyone.
- There may be more flexibility than in larger businesses.
- I will probably have more variety in my job due to the size of the firm.

- Someone else has the responsibility.
- I have a set wage and working hours.

Additionally, in a large business:

- I'll probably have a union to protect my rights.
- Seniority may provide more job security.
- Pension plans and company health plans may provide additional security.
- Opportunity for advancement is greater.
- My job may not have much variety.
- My job will be very inflexible.

SUMMARY

This chapter has attempted to provide some insight into the world of small businesses. Entrepreneurship doesn't satisfy everybody. Many of you, like Jennifer in Figure 1–5, have already had some entrepreneurial experience. Businesspeople have a number of characteristics that are stronger than those of the general population. Included are organization, self-confidence, energy, problem-solving skills, low need for status, a sense of urgency, vision, flexibility, persistence, optimism, independence, emotional stability, commitment, and reasonable risk taking.

Figure 1–5 Many young people get their start in self-employment as babysitters. Jennifer and Jason share a Saturday morning. (Courtesy Susan M. Peters)

Businesses make or produce a product, provide a service, or sell products or services. Self-employment offers an individual the opportunity to take a risk and realize the profit or loss on the decisions and management. Working for others provides a job where others have the responsibilities and set the priorities.

Another chapter will let you analyze your potential for entrepreneurship in greater depth and detail. Entrepreneurship is growing again after decades of shrinkage in numbers.

If you should decide to go into a small business as the result of this course, don't go it alone. Get the help that is available from parents, friends, teachers, and experienced entrepreneurs. In addition, plan carefully and thoroughly. Study the enterprise and consider working for someone else in the business to learn more about it and to gain firsthand experience.

STUDENT ACTIVITIES

1. Develop a list of small businesses in your community that have fewer than five employees. Classify them as to their type of organization, then as to their production service or sales status.
2. Interview a small business owner and ask the following types of questions:
   a. How did you decide on this type of business?
   b. When did you start this business?
   c. What were some of your problems in starting?
   d. Would you advise a young person to go into this type of business today?
   e. Have you had other businesses?
3. Make a list of businesses a student could start.
4. Collect news items for one week. Collect all job openings and items relating to new businesses, and collect all notices of lay-offs or business failures. What is your conclusion about small businesses?

SELF-EVALUATION

Should I Work for Me?

On a separate sheet, write T for true statements and F for false statements.
1. Cash flow refers to the indebtedness of a business.
2. Entrepreneurs take many large risks.
3. Planning is a good way to reduce risk in a new business venture.
4. Collateral is what is pledged to secure a loan.
5. Entrepreneurs have some characteristics that seem to be more highly developed than they are in non-entrepreneurs.
6. Persistence is not a characteristic of an entrepreneur.
7. Entrepreneurs are typically pessimists.

For the following multiple choice questions, write the letter representing the best answer to the stem question.

8. An entrepreneur is most likely to
   a. bet on the horse with 10 to 1 odds at the racetrack.
   b. take the 3 to 1 odds.
   c. back the favorite.
   d. refuse the bet entirely.

9. Entrepreneurs are usually the people who
   a. like working set hours.
   b. like working for someone else.
   c. are independent and believe they can do the job better.
   d. like risky investments.

10. A sole proprietorship is a business that has
    a. multiple owners.
    b. partners.
    c. stock shares.
    d. a single owner.

11. A partnership is a type of business ownership that is represented by
    a. limited liability.
    b. shared capital.
    c. stock shares.
    d. unlimited liability.

12. Company innovators represent the person who is entrepreneurial
    a. within a sole proprietorship.
    b. in a partnership.
    c. in a stock company.
    d. in a limited partnership.

13. Franchise businesses represent the concept of
    a. look-alike businesses.
    b. developers.
    c. innovators.
    d. scalpers.

14. Businesses are classified frequently on their
    a. type of product or service.
    b. number of owners.
    c. sales force.
    d. sales volume.

15. The longer a business survives, the better its
    a. sales volume.
    b. chance for sale.
    c. chance for continued survival.
    d. financial management.

16. Most businesses fail because of
    a. inadequate planning.
    b. good management.
    c. economic cycles.
    d. inadequate funding.
17. The most important advantage to self-employment for the entrepreneur is
    a. freedom to set the time schedule.
    b. the opportunity to create something.
    c. the right to manage other people.
    d. freedom from unions.

CHAPTER 2

How Do I Start A Business Plan?

OBJECTIVE

To identify the parts of a *business plan* and select a type of business to plan.

COMPETENCIES TO DEVELOP

After completing this chapter you will be able to:

1. Identify correctly the major parts of a business plan.
2. Name the four major risks in starting a new business venture.
3. Correctly define 8 of the 10 business terms contained in the chapter.
4. Name three important reasons for preplanning.

TERMS TO KNOW

Advertising	Documentation
Assets	Enterprise
Business expenses	Expertise
Business plan	Federal law
Balance sheet	Financial plan
Business regulation	Fixed assets
Capital	Frivolous items
Cash flow	Incentives
Cash flow analysis	Income statement
Chamber of Commerce	Interest
Competition	Intermediate assets
Corporate managers	Intermediate liabilities
Creditor	Lending committee
Current assets	Liabilities
Current liabilities	Location

Long term liabilities	Regulations (federal, state, local)
Management	Reputation
Market (marketing)	Self-discipline
Marketing plan	Top 10%
Planning commission	Venture capital
Preplanning	Word of mouth
Prime rate	Zoning restrictions

INTRODUCTION

How do I start a business venture? I develop a business plan. All or most business ventures start from an idea—an idea that the entrepreneur can do something new or better. The entrepreneur believes that this idea can be used to get something the entrepreneur wants.

By this time, you have decided that a personal business venture can be exciting and profitable. You've done some self-evaluating. You may have a little more faith in your own creativity, and in your personal sense of adventure. You appreciate the amount of *self-discipline* and organizational skills that an entrepreneur must possess in order to build a successful business. The big question is, "Where do I go from here?" The answer is that it's time to develop a business plan. Figure 2–1 gives a quick picture of the development of a business venture. Figure 2–2 show one self-employed entrepreneur.

An Idea
+
Study of Competition
+
A Marketing Plan
+
Your Resources
+
Good Management
+
A Good Location
+
A Sound Financial Plan
Showing Cash Flow Projections
=
A Successful Venture
Leading to
Satisfaction and
$$ *Profit* $$
_____
(The Bottom Line)

Figure 2–1 The big idea flow chart. (Courtesy The Three Entrepreneurs)

Figure 2-2   Many people reach the goal of self-employment by operating a home-based business. Mike does accounting at his computer. (Courtesy Administrative Business Consultants, Inc.)

In this chapter we will examine what a business plan is, and in later chapters we will study each phase of a plan in greater detail.

CASE STUDIES

The Case of the Undesirable Transfer

Ted immigrated from Sweden as a young man because of the business opportunities in the United States. He found a good job with a large extermination company. He enjoyed his job as a route salesman. He married, bought a house, and felt happily settled in his new community. Ted worked hard at his sales, and was consistently rated in the *top 10%* of the company's route sales representatives. He was satisfied with his salary and commissions. When his company needed a sales manager in another state, Ted was told to prepare to transfer.

Ted liked working directly with his customers, and didn't want an office job managing a large number of salespeople. Neither he nor his wife, Mary Ruth, wanted to move to a new community. In addition, housing markets were down, and Ted would take a financial loss on the house he had purchased. He knew, too, that living expenses would be higher in the larger city, and would consume most of the raise his promotion would bring.

Ted wasn't given the opportunity to refuse the transfer. If he wanted to continue working for this company, he would have to move. It was Mary Ruth who suggested that Ted set up his own small extermination company. He knew the business well after his experience with the large company. Although they would have to live on a tight budget for a year or two, Mary Ruth's salary as a registered nurse would pay the bills. He could set up an office in his basement, hire a secretary and an employee to work in the field, and continue to devote his time to direct sales. He would need to invest some of their savings to get the business started.

Ted and Mary Ruth decided that the risks involved in starting a new *enterprise* were worth taking in order for Ted to continue working at the kind of job he enjoyed.

The Case of the Business Plan

Ted knew how much money they would need to get the business off the ground and to keep it running for the first couple of years. Ted's 4 1/2 years with the company and actual experience with costs and service fees provided the information needed. Although they had a good idea of how much *capital* they would need to borrow from the bank, they didn't write everything down. They didn't create a specific business plan.

Ted had a general idea of the expenses he would need to cover. He understood his product, and he had a good idea about the competition he would be facing. But he didn't have a detailed *financial plan* when he talked to the lending officer at the bank. The lender wanted to know exactly how much money Ted needed for what expenses, and when he wanted it. The lender also wanted Ted's projection of when he would begin to make a profit, and what sort of repayment schedule he would be able to follow. Without that sort of detailed information, the bank felt that Ted's business venture was not sufficiently planned. His loan application was turned down.

Needless to say, Ted was extremely disappointed when he received the rejection letter from the bank. He really believed he could make his own business a success, and he thought he had convinced the lending officer of that.

Ted first thought of where else he might turn to find venture capital. Then he realized that if the bank had rejected his loan because of a poor financial plan, another investor would probably feel the same way. Then Ted realized that if he could put together a solid financial plan showing when and for what he would need the financing, and how and when he planned to repay the loan, then perhaps the bank would reconsider his application.

Ted spent 15 to 20 hours for the next 12 weeks developing his financial plan. He listed everything from equipment and *advertising* expenses to approximate tax payments. He projected a sales quota, and from that he developed an *income statement*. He worked up a *cash flow analysis* for the next two years. It projected *cash flow* based on sales and costs on a realistic basis. Then he made another appointment with the lending officer at the bank.

The lending officer was impressed with the work Ted had put into his financial plan. He was also impressed with Ted's belief in himself and with his persistence.

Ted's extra work paid off. The next week he received a phone call from the banker informing him that his loan application had been approved by the *lending committee*, and that he had a line of credit for his new business venture. Bonnie Jones, loan officer, congratulated Ted and Mary Ruth on their successful application.

What the Cases Tell Us

- Sometimes change is forced by *management* decisions.
- Too many *corporate managers* are more concerned with the company's welfare than with its people.
- Entrepreneurs are willing to risk security to develop their ideas.
- An entrepreneur's family is a powerful support group in starting a new enterprise.
- A good record and reputation are not enough to obtain operating capital.
- Good financial records and *documentation* (a good business plan) will help obtain capital.
- Preplanning is essential in starting a new business.
- Experience in the business helps support the loan application.
- Bankers are not risk takers; they try to make sound loans. They are responsible for people's savings and are regulated by law.

A BUSINESS PLAN SERVES AS A ROAD MAP

Why Create a Business Plan?

We mentioned the organizational skills that an entrepreneur must have. This is where the organization really begins. It's time to put your thoughts on paper and to take a hard look at the chance that your business has to succeed. A business plan acts as a road map (see Figure 2–3)—

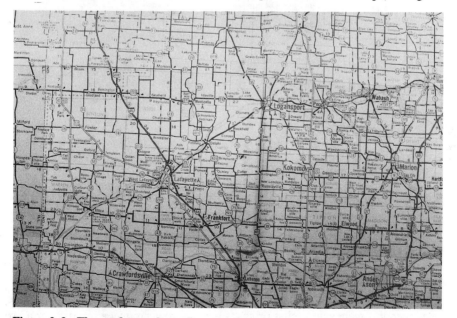

Figure 2–3   The road map shows the route for an auto trip. The business plan guides the way into entrepreneurship from start up to profit. (Courtesy The Three Entrepreneurs)

it will help you assess your personal capabilities. You'll begin to see the amount of "mind," "muscle," and "money" resources that are required for a new business venture. You weighed the advantages and disadvantages of owning your own business. Do you have the commitment you'll need? Are you prepared to weather some rough times? A business plan will help you realistically identify where and when to expect the greatest challenges. It will show you what resources you'll need to get started, what time and energy you'll need to invest, and when you can expect to see a profit. A realistic business plan will also be your most important tool as you try to convince creditors and other interested parties that your business can succeed. The following are steps that you need to think about. Each step includes important questions. Write down the answers to the questions that apply to your business venture. We often use what we have to reach our goals. The Buck family illustrate such use in Figure 2–4.

Figure 2–4   **The Buck family used the resource they had. They remodelled the old barn to fit the swine enterprise. (Courtesy David Buck)**

Your Product or Your Service

Consider the product or service that you've decided on. Can you convince others that your product is as good as you believe it is? Will people want it? Will they spend their money on it? Are they spending money on similar products now? Is your product a necessity or a luxury? What is the economy like, and what is it expected to do? During a recession, people are less likely to spend their money on *frivolous* or luxury *items*. Is your product affordable?

The Competition

Will you be facing *competition*? If so, can you offer a better quality or less expensive product in order to capture a profitable share of the *market*? Can you target your product or service to a particular area in which you are able to specialize?

Your Market

Consider the potential growth of your company. Do you want to keep it small enough to manage from your own home? (The January 1986 *Reader's Digest* gives many examples.) Do you hope to expand some day to regional or national sales? Is it possible to saturate the local market with your product so that you'll no longer fill a need? Is it possible that your product or service will become technically obsolete within a few years?

There are resources to help you learn more about your market. The *Chamber of Commerce* can give you information about products and services produced locally. The library can be a source for information on a national scale. Many states and cities have industrial planning agencies which can also be a source of information.

Resources

These are the questions about resources for which you will need answers. What resources will you need? Who will supply the materials you'll use? Are your supplies easily accessible, or will you be needing limited resources that must be transported long distances? Check with potential suppliers. Can they meet your needs? Can they provide timely service? A product is not ready to sell until all of its parts are in place. How much space will you need? How much space might you need for expansion? What equipment or machinery will you need to purchase? How many people will you need to hire?

Business Regulations

There are many *federal, state, and local regulations* for businesses. Your local *planning commission* may specify waste disposal, for example. *Zoning restrictions* regulate the *location* of businesses. Individual states have different laws dealing with insurance requirements. *Federal law* regulates the working environment you provide your employees and the tax structure of your business. While some *business regulations* may seem simple, others are very complex. At some point your business will need legal advice. This is the time to consult a lawyer. A lawyer will inform you of the regulations that you must comply with. He can also give you tax advice. While the expense of a lawyer may seem unnecessary at first glance, remember that you are paying for knowledge and *expertise* that you don't have. At this time be sure to use your support team. A very good source for information is the Small Business Administration (SBA). This is a government agency with regional offices, and is listed in the telephone directory. This agency studies small businesses in the United States, and has information on many subjects from regulations to competition concerning businesses just like the one you're planning.

A Marketing Plan

You may create a tremendous product, but you won't see a profit from your work if nobody knows about it. Therefore, a *marketing plan* is needed. You have to inform the public about your business, and you have to convince people that they want your product or service. You want to generate sales and create customers. This is done by advertising. It is said that *word of mouth* is the best advertisement. This may be true, and you want to build an honest and reliable *reputation* for your company. Advertising by word of mouth is a slow process. You need to introduce your product enthusiastically and quickly. Look around at the promotions that other businesses invest in. They advertise on TV and radio. They buy space in newspapers or mail fliers to local residences. They plaster their products on billboards, and write letters to potential customers. All of this costs money. In fact, advertising is a $50 billion industry! Businesses often pay for researchers to study the market in order to develop the most effective advertising for the product. They become experts at knowing what produces a profit and what has little effect on the public.

Who do you think will buy your product? How do you reach them? How can you catch their attention? What forms of advertising appeal to you and to customers? What does it cost? Call the newspaper and the radio stations. Find out what their advertising costs. This is a time to put your creativity to work again.

How much of your budget do you want to invest in advertising? What do you believe will be most effective for the money you've decided to spend on promotions?

Your Location

You might be amazed at the money fast-food chains invest to research possible locations before opening a new store. They know that location can be the key to the success or failure of a new business. You can benefit from their experiences. If your business will depend upon customers coming to your location, you will want to find a place that is easily accessible and highly visible. If you will be offering a mail-order service, or if you'll be traveling to your customer, perhaps a less conspicuous location would be more appropriate. If you're looking for an existing building to convert to your needs, how much floor space do you need? How much will you need a year from now? Ten years from now? Do you need a showroom? Office space? Unfortunately, one of the largest factors in the cost of business space is location. You may find that your budget requires you to make compromises for the present time. If so, don't rule out the possibility of changing location in the future. And don't put a lot of money into frills for space that you may use only temporarily.

Your Employees

Many entrepreneurs don't hire employees when they start out on their business venture. As the model of entrepreneurial resources shows, the entrepreneur is usually the "mind," "muscle," and "money" behind the business. You may discover, though, that you will need to hire a couple of employees, if not in the beginning, then later as your business grows.

The decision of how many employees you'll need to hire does not end your responsibilities as a manager. What kind of employees do you want to hire? Will they need special skills? Do you want to hire people with technical training, or can you train them to do the job yourself? Can you convince a prospective employee that your business will succeed? Your business success may depend on persuading others to join you. You can always hire somebody who needs to work, but you need the best. Your business plan can help here.

What kind of *incentives* can you offer that will convince someone that he or she wants to work for you? Will you offer a salary or an hourly wage? What kind of benefits can you give an employee? Will you offer sick leave? Paid vacations? A pension plan? Insurance? How will you review their performance? What kind of severance plan will you have? Can they expect promotions? Raises? Remember, when you do find the right employees, they will need time to terminate their present jobs before they can begin working for you.

Management Organization

How will you organize your business? Will it be a simple individual ownership where you oversee one or two employees, or will you have several employees working in different departments? In the latter case, you will probably find that you'll need to appoint a supervisor for each department. Large businesses have management levels, and the management of the business is structured in a hierarchy.

The Financial Plan

Your financial plan will probably be the most difficult stage of your planning, but it is the most important. Not only do you need to figure very closely the total cost of getting your business off the ground, but you need to have a very good idea of how much money it will take in. You need to know when to expect the expenses and the income. During the startup period of your business, you will be spending more than you are making. Cash flow projections are vital!

An Income Statement.   An income statement is sometimes called a profit and loss statement (a P&L). It reports the financial results of your business operation by measuring your profits against your expenses. If, at any certain time, your profits are greater than your expenses, your income statement will show a net profit. If your expenses are greater than your profits, your income statement will show a net loss. As your business gets underway, you will create actual income statements. For now, you want to project what you think your profits and expenses will be at specific times. It's a good idea to project a monthly income statement for your first year, a quarterly statement for the second year, and an annual statement for the third year.

A Balance Sheet.   A *balance sheet* shows the *assets* and *liabilities* of your business at a particular time. It lists all of a firm's assets (those things that it owns) and all of its liabilities (the money it owes). It assigns a monetary value to each asset and each liability, and then totals both groups. The liabilities are subtracted from the assets to produce a net worth. Assets and liabilities are usually divided into three groups. The Formula:

$$\text{Assets} - \text{Liabilities} = \text{Net Worth}$$

Assets include:
- Current Assets—*Current assets* include cash on hand, money in checking and savings accounts, marketable stocks and bonds, and money that will be paid to you within the next year.

- Intermediate Assets—*Intermediate assets* include payments that will be made to you after a year's time. They also include the value of any machinery, equipment, or inventory that your company owns. The value of a retirement fund and the value of nonmarketable stocks and bonds would also be included as intermediate assets.
- Fixed Assets—*Fixed assets* are items such as your building, rental properties, land, and long-term contracts.

Liabilities include:

- Current Liabilities—*Current liabilities* are those bills that you must pay within the next year. This includes loan payments and money owed to *creditors*.
- Intermediate Liabilities—*Intermediate liabilities* include any payment that is due beyond the next year, such as the remaining principle on a loan.
- Long-Term Liabilities—*Long-term liabilities* include the money that you owe on your fixed assets, such as the remaining principle to be paid for a mortgage. Do not include that portion that will be paid within a year, as that has already been listed as a current liability.

The Cash Flow Statement.   The cash flow statement actually measures the flow of a company's cash over a period of time. We mentioned the beginning period of a business when it will be spending more money than it is making. *Business expenses* and profits are not always regular. A garden center makes more money in the summer. A farmer pays the bills for his corn crop in the spring, but doesn't reap the profits until in the fall. The expense for his winter wheat crop comes in the fall, while he sees income from that crop the following spring. You can begin to see how many businesses have a cash flow cycle that changes frequently. It's important for a business manager to know when to expect his greatest expenses, and when the profits will be coming in. There may be certain times of the year when the business may need to borrow more money. The interest it will pay for that money will continue to be an additional expense until the profits can cover the loan.

A cash flow chart usually details the amount of income and expenses a business has each month for 12 months. For now, you will want to use it to project the cash flow of your company for the first year. As your business begins to operate, you'll want to compare the actual cash flow to your projection. Cash flow analysis will continue to be a part of your business management for the life of your business.

Financing Your Business

Following these steps, you should have a good idea of how much money you will need to get your business started and to keep it running until your income begins to cover your expenses. Few entrepreneurs have the personal capital to start up their own businesses. They use other people's money. Keep in mind that there is a price tag attached to the use of someone else's money. That price is called *interest*. The interest you'll pay will vary with the source of the capital. For example, interest rates on a car loan at the commercial banks were roughly 14%, while credit unions were charging 11%. Credit card interest rates were as much as 21% when the prime rate was about 9%.

There are different sources for capital. Perhaps someone in your family or your circle of friends has savings that they would consider investing in your business. It's possible to work out a mutually beneficial deal. You can offer them a higher interest rate than they are currently receiving for their money and they can offer you financing at a cheaper rate than a local lending institution.

There are people who invest in new businesses because of attractive interest rates. The money they invest is called *venture capital*. Talk with business people that you are acquainted with. If you believe in your business idea, you can convince others to believe in it, too.

Other alternatives are your local lending institutions. They are in business to make a profit from money they lend to people like you. You'll need to convince them that you can make your business a success. You can do this with a carefully researched and developed business plan.

Remember, your financing may not all come from one source. The bank may be willing to lend you a portion of the amount you need if you have venture capital from another source.

Selling Your Business Plan

When you talk to people about your business, you are actually selling yourself and your ideas. Just as in any sales job, your attitude and your appearance are important. You want to appear confident. Don't be afraid to let your enthusiasm show. You don't want to claim that your venture is risk free, but you do want to be able to show that you understand the risks and that you believe your venture will be successful. It might be all right to approach Aunt Mary in your blue jeans, but dress conservatively if you're headed to the bank. Be well prepared. Have a business plan on paper that shows the research you've put into your proposal. Have your proposal neatly typed. Include the following sections.

A Summary.   Begin with an overview of your objectives. What sort of business are you proposing? What qualifications do you have that will contribute to the success of the business?

The Background.   How did you come up with your idea? Why do you see a need for this sort of business? Why do you feel this is the right time?

Your Analysis.   Why are you convinced that there will be a market for your product or service? You've studied your competition—specify that in your proposal. Who is it? Why do you believe you can corner a share of their market? What advantages can you offer over existing competition?

Your Marketing Plan.   What is your market? Who will buy your product? How will you advertise it? How much will advertising cost?

Your Financial Plan.   This is when you talk about the amount of capital you'll need. You'll want to be very specific. How did you arrive at that figure? How much will it take to buy the equipment you'll need? How much for rent? Salaries? Insurance? Supplies? How much will it take to keep your company in operation before it becomes profitable? How do you intend to repay your loan? Will you want to borrow capital for six months? Two years? A lender will be impressed if you know the answers to these questions. He'll probably send you back out the door if you don't. Include copies of your projected income statement, your balance sheet, and most importantly, your cash flow study.

Your Location.   What location have you chosen for your business? Why is this the best alternative? Will you operate your business out of your home? Are you renting or leasing space? Are you buying and renovating an existing building?

Your Time Schedule.   Have an idea of the time you will need at each step in starting your business. When will you make purchases? When will you hire employees? When do you expect to begin production or services?

STUDENT ACTIVITIES

1. Search economic texts for business plans and bring them to class to discuss how they are introduced.
2. Present copies of a business plan from one of the references to provide a discussion on business plans.
3. Start an outline of a business plan for the business venture that is being considered.

SELF-EVALUATION

The Business Plan Quiz

1. List the major parts of a good business plan.
2. What are the major risks in starting a new business venture?
3. Give a brief definition for each of the following terms:
    a. Assets
    b. Liabilities
    c. Cash flow
    d. Balance sheets
    e. Business plan
    f. Competition
    g. Location
    h. Marketing plan
    i. Regulations
    j. Venture capital
4. Why is preplanning important?

CHAPTER 3

How Do I Set Personal and Business Goals?

OBJECTIVE

To develop the ability to set goals for yourself or a business venture.

COMPETENCIES TO BE DEVELOPED

After completing this chapter you will able to:

1. Describe a minimum of 10 motivations for an entrepreneur.
2. Identify three kinds of goals based on the time needed for completion.
3. Inventory the time available for an entrepreneurship activity.
4. Describe a successful goal-setting procedure.

TERMS TO KNOW

Behavior	Emotional motivation
Human motivation	Incentive
Objectives	Office politics
Obsession	Physical motivation
Priorities	Procrastinators

INTRODUCTION

You probably know people who always have a big plan but whose ideas never seem to get beyond the planning stage. Suddenly last week's idea is replaced by a new one. These people are "dreamers." They spend their lives thinking about what they might do someday, but they never actually do much of anything. On the other hand, you probably know people who set definite goals for themselves and pursue those goals with determination until their idea becomes a reality. These people are "doers." Are you a "dreamer" or a "doer"?

There are some questions about goal setting that we should consider.

- What motivates people to set and pursue goals?
- What is the difference between a dream and a realistic goal?
- How does this apply to you as you start a new business venture?

Opportunity alone does not pave the way to success. Let's look at a couple of cases in which motivation and opportunity differ.

CASE STUDIES

The Case of the Sudden Windfall

After Jim Allen graduated from high school, he went to work for a grain elevator. He earned enough that he could put some money aside each month for a new car, which had been his dream since high school. When Jim was 22, he inherited $20,000 from his grandmother's estate. The grain elevator that Jim worked for was looking for capital in order to build three new grain bins. Jim's father, Ralph Allen, advised Jim to approach the owner of the elevator with an offer to invest his $20,000 in return for a share in the ownership of the business. Mr. Allen saw that Jim had an opportunity not only to invest his inheritance in a business enterprise, but also to create a better job opportunity for himself. Mr. Allen believed that Jim might even be able to become a partner in the business someday.

Jim, however, saw things in a different light. He'd been eyeing a shiny, new Camaro that had been sitting on the lot of a Chevrolet dealer for about three weeks. He didn't see any point in sinking his money in a business venture when he could buy what he wanted now. His goal was similar to Joe's, Figure 3–1.

In spite of his father's advice, Jim decided to buy the car. He drove it proudly to work at the elevator every day, where he continued working for a series of small raises for the next several years.

The Case of Winning Determination

For as long as he could remember, Eric had wanted to fly airplanes. Perhaps his *obsession* began when he was 5 and his family had flown on a big jet to Florida for a winter vacation. When Eric was in the sixth grade, his family moved to a new home that was about 3 miles from a small, privately owned airport. That summer, Eric rode his bike to the airport almost every afternoon to watch the planes. It wasn't long before his interest caught the attention of the men who worked in the hangar. They let Eric watch them work, and patiently answered the questions that Eric always seemed to have. By the next summer, he was running errands for them and helping with minor repair work on some of the planes.

During his junior year in high school, Eric had an idea. He offered to exchange his work in the shop at the airport for flight instruction. Two years later, Eric flew his first solo flight. He continued until he had enough money saved to buy his first airplane.

Today, Eric has an aerial photography business. He uses infrared photography to help area farmers locate the different types of soil in their fields. By studying the information that Eric

Figure 3.1   Joe's goal was to own a one-of-a-kind auto. This is the goal he achieved. (Courtesy Larry Hart)

supplies to them, the farmers can use fertilizers and other soil additives more efficiently and save money. Eric developed his dream into a realistic goal, which in turn became a successful business.

Examining the Cases

We might call these cases a study in opportunity and motivation. Consider these questions.

- What goals did Jim set for himself?
- What were Eric's goals?
- What opportunities did Jim have?
- What opportunities did Eric have?

Jim and Eric both had goals. Jim's was to own a new car and Eric's was to fly airplanes. Jim's goal, however, was a very short-term goal. It was also short-sighted. Jim failed to see, even with his father's advice, that in years to come he might want other things that a better job could provide. He saw the money he inherited from his grandmother as his opportunity to reach his present goal, which was to own a new car. He should have been thinking about longer term goals. Jim missed the opportunity to improve his job.

Eric's goal was to fly. His opportunity lay in the fact that there was an airport he could reach by bicycle. Other than that, he created his own opportunity with his persistence and determination. Eric didn't stop when he reached his goal of flying an airplane. He expanded his goals, first into owning his own plane, and then into developing a business for himself. Not only could he continue his flying, but he could also provide a service for which farmers were willing to pay. Eric turned something that he enjoyed into a profitable business enterprise.

Sometimes unexpected opportunities provide motivation for us. Think of the person who has suddenly lost a job because of company layoffs. It's hard to think of that situation as an opportunity, but the need for employment will motivate that person to search for a new job. New possibilities arise that a steadily employed person would never take time to consider. Frequently, people find even better jobs because of their search. Many new businesses are started for just this reason. People think about starting their own business for years, but aren't bold enough to take the big step until something happens that returns them to the job market. Sometimes need supplies the motivation for an entrepreneur.

GOAL SETTING

Motivation

Motivation is defined as an incentive for *behavior*—an emotion, a desire, or a physical need that causes us to act.

When we're hungry, we eat. When a storm suddenly comes up, we take shelter. Those are examples of *physical motivation*. When we're upset, or we need help with a problem, we talk to someone we trust, hoping to find some sympathy or advice. Those are examples of *emotional motivation*. The causes of *human motivation* become complex, and we won't go into the many theories that are available. The point is to make you think about what motivates you. What turns your crank?

First, think about your own expectations. Consider the importance of the following ideas. The list can help you decide what you want to accomplish with your business.

- Money
- Material things
- Recognition
- Authority
- Respect
- Personal satisfaction
- A sense of accomplishment

One characteristic of an entrepreneur is that this person works to fulfill his or her expectations, rather than the expectations of other people. Ronnie set his own expectations, Figure 3–2. You can probably see how important this is from your schoolwork. Aren't there some studies which seem to come easily because you're interested in them? At the same time, you probably have classes that you study harder for because of the grade you'll receive on your report card. Think about the reasons that you do the things you do. Are you working to please yourself? Your parents? Your teachers? Your friends? People often perform according to what they believe others expect from them. That's terrific when it gets your room straightened up every morning, or gets you to study extra hard for a big exam. But entrepreneurs have a personal drive to succeed. They know that they can expect some rough times in their venture, and it is their personal drive that helps them through these difficulties. It is easier to give up when you're doing something because of someone else's expectations. The important thing is to be able to define your own realistic goals. Then, pursue these goals because *you* want to, not because you're concerned about what someone else expects of you.

Figure 3–2   Ronnie does auto, truck and tractor repair in his business. He developed his business plan in the Western High School entrepreneurship class. (Courtesy Ronnie Orem)

Motivation for the Entrepreneur.   In Chapter 1, we talked about some of the advantages of working for yourself. This is a good time to review the advantages that are the motivation behind entrepreneurship.

- You work for yourself. You're independent. You're the boss.
- What you earn above expenses is all yours.
- You make the decisions that affect your life.
- You can test your own ideas.
- You set your own working hours.
- You don't pay union dues.
- You'll have a variety of tasks.
- You won't have to worry about *office politics*.
- You set your own prices according to the market.

- You determine your production schedule and control your inventory.
- You determine your product or service.
- You control the quality of your product and the reputation of your business.
- You solve the problems.
- You hire, train, and fire your employees.
- You control the advertising and business image.
- You set the company policy.
- You get to "turn out the lights" or close for the day.

What are Goals?

A goal is something we strive to achieve. It's an end toward which we direct our energy. Goals give direction to our lives. Without them, our existence would be pretty dull.

Why Should I Set Goals for Myself?   We set goals in order to plan for the future. We decide what we want to do, or to be, or to have, and then we begin to work toward those goals. We can all express general goals. We want to be happy. We want to have good health. We want to live comfortably. It's harder to define specific goals. Many people spend a good part of their lives traveling in circles, not sure where they're headed or how to get there. Figure 3-3 illustrates a young entrepreneur using the resources he has to begin working toward his goal.

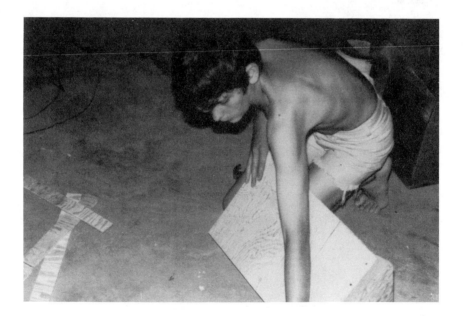

Figure 3-3   Many young entrepreneurs use resources on hand during start-up. Dan builds berry picking boxes. (Courtesy Barbara Doster)

Pursuing goals is like following a road map. If you want to get from Kokomo, Indiana to Bangor, Maine, you don't sit in your chair and wish you were there. You chart a course on a map following the best roads and quickest route, and then you get started. You divide your route from intersection to intersection, and take it one stretch at a time. Ultimately you reach your destination. You can create a time line for the direction you want your life to take just the way you follow a road map on a journey. You just plot out every step on the way to your goals, and then follow your plan, one step at a time, until you achieve them.

By setting specific goals for ourselves, we can plan the course we need to follow in order to attain them. Goals make life exciting and give us a challenge. More important than that, when we define goals, we are more likely to achieve them.

Why Doesn't Everyone Set Goals?   If goals are so important, why doesn't every person establish goals? There are several reasons. First of all, goal setting is a skill. Some people can't clearly define what it is that they want to do. They have a vague idea of what they would like, and hope that someday things will fall into place for them. Another reason some people avoid setting specific goals for themselves is that they're afraid of failure or they have a poor sense of self-esteem. Sometimes people don't feel that they deserve the best in life, and so they don't pursue the things they want. Finally, some people simply don't think about setting goals. They do things the same way that others have always done them. When you decide to pursue a goal, use your creativity and trust your own judgment. While it's important to learn from the experience of others, it's also important not to waste time by doing things that work for someone else, but that may not be right for you.

How Do I Set Goals?   There are some general rules which can help you set goals for yourself.

Write Down Your Goals.   The best place to start as you begin to think about the goals you want to reach is by writing them down. Make a list of everything that you would like to accomplish. At this point it's all right to be a little bit unrealistic. Put down everything that comes to your mind, from always keeping your car clean, to taking a trip to Europe. (Everyone taking this course will want to include starting a business.) Think about what you want to do tomorrow, 10 years from now, a lifetime from now. Your personal goals may involve establishing a family, planning vacations, owning your own home, or building a retirement fund. You may have educational goals, financial goals, career goals, and social goals. Try to cover all areas of your life.

Organize your Goals.   You will have both specific goals that you want to reach in a few months, and goals that you will work toward for many years. In between these, you will have goals that will take a couple of years to reach. Arrange your goals according to these three groups.

- *Immediate goals.*   Immediate goals are the goals that you would like to accomplish within the next year. For you as a student, these might include graduating from high school or making your car payments. For you as an entrepreneur, these immediate goals would probably include the first steps you'll need to take to get your business started.

- *Short-term goals.*   Short-term goals include the things you want to accomplish in approximately the next 5 years. They often include the steps you need to take to build toward your long-term goals. For instance, you might include getting your own apartment as a personal goal. For an entrepreneur, business expansion or marketing perfection would fit between immediate and long-term goals.

- *Long-term goals.*   Long-term goals are the ones you intend to work toward for many years. They summarize everything that you plan to work for in your personal and business life. Long-term goals might include saving enough money to send your children to college, or building a pension plan for a comfortable retirement. Long-term business goals give you an idea of what you want to do with your business many years from now. Perhaps you want it to grow big enough that it will become a corporation. On the other hand, you may want it to remain a small and solid family venture.

Evaluate Your List.   Cross out those things that aren't really important to you. Then take a hard look at the goals that aren't really practical. Some of your goals may conflict. If you would like to run a ski lodge in Colorado, but would also like to live in Florida, then you need to decide which of those goals is more important to you.

You will need to consider other people as you establish your goals. If your family will be affected by your decisions, then it's important to consider the sacrifices that you may ask of them when you reach for your goals. Perhaps you'll want to adjust your goals in order to benefit others.

What you should have on your paper is a list of realistic and attainable goals. You're now ready to think about what it takes to reach for those goals.

How Do I Reach My Goals?

Managing Your Time.   The first step toward reaching your goals is to learn to manage your time. In order to do this, you must understand the way that you spend your time. Everyone has 24 hours to spend in each day and 365 days to spend in each year. Some people accomplish a great deal using the amount of time that they have. Others do not. Consider the 24 hours that you have each day. How do you use them?

Figure 3–4 shows the activities an average high school student fits into a day. We've divided the day into two halves, A.M. and P.M. The shaded portion of each graph demonstrates the amount of time that the average student has to pursue unscheduled activities. Multiply that by 5 days a week, add the 48 hours each weekend, and you'll see that there is a surprising amount of time in which much can be accomplished.

Everyone has committed a portion of time to certain activities. As a student, you spend a specified number of hours in school. You need a certain number of hours each night to sleep. A portion of your day is used to eat meals. You spend time getting ready for school and traveling to and from school. Some of your time is set aside for homework and chores. But there is time each day which is flexible. Sometimes we call this "leisure time." Time management involves the way we organize that extra time. During our school years, we use a large portion of that time for school-related activities, such as sports events. Parents often devote much of their extra time to their families. Entrepreneurs quickly discover that they devote the greatest amount of their leisure time to getting their business established.

There are three techniques that lead to successful time management.

- *Avoid procrastinating. Procrastinators* are especially good at putting off anything that doesn't have to be done immediately. Good time managers use all the time they have to progress toward their goals.

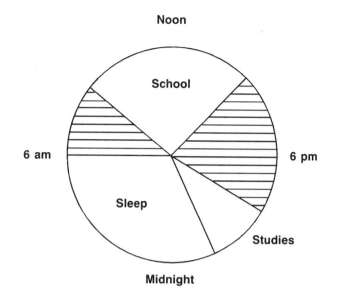

Noon

School

6 am

6 pm

Sleep

Studies

Midnight

Figure 3–4 Inventory of time available. (Courtesy The Three Entrepreneurs)

- *Judge your time.* The second technique you'll want to develop is the ability to judge the time you'll need to accomplish a task. With experience, you'll be able to have a very accurate idea of how much time different activities require. This allows you to use the third time management technique.
- *Schedule your time.* Be realistic. You will not perform a task well if you don't allow enough time to complete it properly. On the other hand, don't budget so much time for a project that you find yourself wasting time between appointments or projects. This leads us to a discussion about *priorities* and the way we use our time.

Establishing Priorities. Over the course of a lifetime, the amount of extra time that we have to spend adds up to an extremely large total. How many people do you know who spend an average of 2 hours an evening watching television?

$$365 \text{ days a year}$$
$$\underline{\text{x}\quad 2 \text{ hours a day}}$$
$$730 \text{ hours a year}$$

$$\frac{730}{24} = 30.42 \text{ days a year}$$

The person who watches television only 2 hours an evening is spending one whole month a year, 24 hours a day, in front of the television set. That's over 91 working days. If that person continued to watch TV 2 hours a day for 75 years, he or she will have put the equivalent of over 6 years worth of working days into time spent watching TV.

2 hours a day

x 75 years
_____

6 years of working days

That amounts to a large chunk of lifetime. You can begin to see how valuable your time is. Put it to work for you. Use it to reach your goals.

The point of this demonstration of how time adds up was to show the importance of establishing your priorities. Now go back to your list of goals. Some of them will be more important to you than others. Using a scale of 1 to 10, rate each goal on your list according to its importance, giving those with the highest priority a 10. Rate the others, comparing them to your most important *objectives*. When you begin to work toward your goals, begin with the ones having the highest priority. When you've done all that you can toward reaching those goals, move on to the goals with the second highest priority. By establishing your priorities, you will be using your time not to work harder, but to work smarter.

Breaking Goals into Manageable Units.   The best goals are the big ones. They're the ones that you set your sights on, the exciting ones. The long-term goals help you overcome the daily frustrations which you will encounter as you work on your immediate goals. But big goals cannot be accomplished in a day. If you break them into manageable units, which become immediate goals, you can see your progress. While continuing to work for the larger goal, you earn a sense of accomplishment that is encouraging. Think about how you can work toward your goal on a daily basis, and then strive to accomplish that much each day.

The Federal Bureau of Investigation (the FBI) has a goal-setting process for its agents from which we can benefit. Of course, the FBI would like to caputre all of the criminals in the United States. That's a long-term goal. In order to make that goal both realistic and manageable, they publish a list of ''Ten Most Wanted'' criminals. Each agent has a copy of that list and each does his or her part to apprehend those criminals. Each day the list is updated. New names are added as others are deleted.

The names on the list comprise the FBI's immediate goals. The names waiting to be added could be considered the FBI's short-term goals. It's a method that works. It helps the FBI work toward its long-term goal. You can make the same technique work for you. Make a list of the 10 things you would like to accomplish today. Strive to cross each item off the list by the end of the day. Tomorrow make a fresh start with a new list. By the end of the week you will surprise yourself with what you've accomplished. You'll have a greater sense of confidence in what you can accomplish when you work at it. If you make a habit of daily goal setting, and multiply that feeling of accomplishment by years of meeting daily goals, you'll be well on your way to accomplishing your long-term goals.

You'll want to have a good understanding of your personal goals before you try to set goals for your business. Do you want to live in a small town or a large city? Do you want to work with a few people or do you see yourself managing a large crew? Do you want to devote a large portion of each day to family or recreation, or do you envision yourself working long hours each day? As you begin to establish your personal goals, you'll get a better picture of the size and type of business structure you'll want to develop.

Planning and Achieving Business Goals

Now that you've given some thought to your personal goals, it's time to think about your business goals. Every successful businessperson establishes goals and plans the way to achieve those goals. The more specific you can be in planning your business goals, the easier it will be to pursue them. Follow the same steps that you used to plan your personal goals. Write down everything you can think of that pertains to a business goal. Divide those goals into the three categories: immediate, short-term, and long-term. Eliminate the less important, conflicting goals, as well as those that are neither practical nor realistic. Then give each goal in each of the categories a priority rating, and rearrange them accordingly.

An Example of Business Goals

Janice Beech is about to embark on a new business venture. The small town that she lives in does not have a flower shop, and people in her community order flowers from a larger town 12 miles away. Janice has always liked working with flowers, and her friends often compliment the floral arrangements in her home. For two years she has made a hobby of growing flowers in a small greenhouse that she built in her backyard. Recently several friends have asked Janice to make up floral arrangements for gifts. Janice believes that she can build a profitable business with her talent. She prepared a list of goals for her venture.

Immediate Goals

- To draw up a business plan so that she will know exactly how much capital she will need and how and when she will repay it.
- To obtain capital and enlarge her greenhouse, buy a refrigerating unit, and open a small shop.
- To produce enough income so that her business will have an effective cash-flow.
- To stay within a specified debt limit.
- To sell fresh cut floral arrangements and potted plants to people in her community.
- To locate suppliers for vases and other items she'll need.

Short-Term Goals

- To repay her original loan.
- To capitalize expansion. Perhaps include silk floral arrangements and garden plants.
- To provide flowers for weddings. Perhaps to add other floral catering services.
- To have an income that provides a satisfactory profit over expenses.

Long-Term Goals

- To hire and train employees so that she can expand the capacity of her company (and take vacations).
- To broaden her market. Perhaps to price her products so reasonably that she can entice new clients from the larger town.
- To establish financial security.
- To sell the company and retire with a handsome profit.

Janice's Daily Goals.   Janice has a good idea of what she wants her business to do and of what she hopes to accomplish. She's excited by her goals, and she takes the first step by listing 10 things that she will do the next day. Here is Janice's "Ten Most Wanted" list for her first day:

1. Decide on a name for her new business.
2. Visit the flower shop in the town 12 miles away and see exactly what it is charging for its floral arrangements and planters.
3. Make a list of all the kinds of flowers she would need to grow in her greenhouse.
4. Make a list of all the flowers, like roses and orchids, that she would have to purchase wholesale.
5. Call three different wholesalers and request their most recent catalogs.
6. Make a list of possible locations.
7. Talk with her father about her idea to see if he would be interested in investing in her company.
8. Draw up a brochure that she could have printed to distribute to her friends and to local businesses.
9. Visit the library and pick up some books about retailing businesses.
10. Make an appointment to talk with Mrs. Ramey, who opened a new clothing shop three years ago, about her experience with a new business venture.

It's obvious that Janice may not accomplish every item on her list in one day, but she will do her best to cross off as many as possible. The next day, she'll add more items to her list so that she'll have a new list to work from. Soon, Janice will be selling flowers from her own shop.

Consider Your Road Blocks

Remember that you will encounter obstacles as you pursue your goals. Some of them can be foreseen. If you're running your own shop, who will cover for you during emergencies? If a supplier cannot make a delivery, who would serve as an alternative? These are obstacles that you can plan around. If you are prepared for the foreseen obstacles, you'll have more time and energy to deal with the unexpected ones.

Evaluate Potential Conflicts

We discussed the fact that your goals often affect other people. If your business goals will require a greater amount of your time, you'll want to consider how this will affect your commitments, such as your family.

Your goals may conflict with the goals of others. If, for instance, you've been farming with your father for several years, be sure that you understand each other's intentions. He may be preparing to retire and not want to expand the operation, while you may be excited about purchasing new ground or equipment.

Communication is the key to human relations. Be sure that those who will be affected by your goals understand what your goals are.

Prepare for Changes

Circumstances change with your age, your health, and your family obligations, among other things. Your attitude may be very different 20 years from now. Be prepared to re-evaluate your goals from time to time. You may find that they need to be restructured because your priorities change with your values. Don't be afraid to make decisions. Don't feel defeated when a decision turns out to be wrong—sometimes even the best decisions can have bad results. The most successful managers are the ones that can profit from their mistakes and move ahead.

Develop New Work Habits

You've thought about the goals that you want to set for your business. Now it's important to develop a methodological approach to breaking your goals into manageable units, and then pursuing those accomplishments one step at a time. Remember the television illustration earlier in this chapter. If you use your extra time to pursue your goals, you can reach them. We've structured a pattern for you to follow in order to establish and reach your goals. Now it's up to you to make daily goals a habit. You'll be pleased with the results!

STUDENT ACTIVITIES

1. Chart your daily routine and use of weekend time.
2. Have a local entrepreneur discuss goal setting with your class.
3. List your immediate, short-, and long-term goals.
4. Prioritize your goals described above.

SELF-EVALUATION

Goal Setting Test

1. Name the two types of human motivation.
2. List 10 advantages to the entrepreneur in starting a new business.
3. Give two reasons for setting personal goals.
4. What are the three steps in setting goals?
5. Name two daily activities to which you are committed.
6. Name two daily activities you could change.
7. What are three techniques that lead to successful time management?
8. List three personal immediate goals.
9. List three personal short-term goals.
10. List three personal long-term goals.

CHAPTER 4

What Type of Business Organization Should I Choose?

OBJECTIVE

To identify the characteristics of various types of business organizations and to select one for a personal business venture.

COMPETENCIES TO BE DEVELOPED

After completing this chapter you will be able to:

1. Identify four types of business organizations.
2. Identify a franchise.
3. Describe the distinguishing characteristics of the four types of business organizations.
4. Select a type of business organization for a personal business.

TERMS TO KNOW

Appreciation
Credit
Regulations
Compatibility
Start-up Costs

INTRODUCTION

We have discussed types of businesses—production, service, and sales—in an earlier chapter. Each of these may be organized as a sole proprietorship, a partnership, a stock company, or a cooperative. In this chapter we will examine each of these forms and look at some considerations for selecting an appropriate form of organization for your business enterprise.

CASE STUDIES

The Case of the New Management Service

Bob has worked with a group of farmers interested in improving their management decisions through better bookkeeping and farm record keeping. When personal computers became available with suitable software, he saw an opportunity to start a computerized record-keeping service keyed to monthly reports and summaries. He believed he could supply a superior service compared to others in the farm management field. Accordingly, he set out to start a completely new business.

Since his capital was limited, he had to make some decisions on how to finance the new enterprise. One of those decisions involved the type of business organization he would adopt. After you complete the reading of this chapter you will be asked to advise Bob in this area.

The Case of the Silent Partner

Howard was a partner in Exeter Motors and in a new feed and grain business. He owned and managed several farms and did not have the time to take an active role in either of the two businesses named. His contribution was an appropriate share of the capital for each. This is usually considered a silent partnership. Howard did contribute both time and management ideas to the businesses so perhaps we should consider him a "not-so-silent partner." When you have completed the reading of the chapter, determine his legal status.

The Case of the Family Corporation

Jack Hanna has three sons and a daughter. As each reached the age of ability to contribute to the family enterprises, the corporation provided a means of giving them an opportunity to invest in the business and share in proportion to their equity in the business. The original shares were given to each family member. After reading the text, list the advantages to the family members in the corporate type of business organization.

What the Cases Tell Us

- A new idea is a starting point for a business enterprise.
- How the idea is developed and merchandised determines whether or not the business succeeds.
- A superior product at a reasonable price is a good business combination.
- The type of business organization affects how a business is financed.
- Partners are active or silent.
- Silent partners participate in business by finances and sometimes by management decision making. Normally, silent partners are not active in the business.
- A corporation allows for many shares or degrees of ownership.
- The family corporation enables all members to have a share of ownership.

TYPES OF BUSINESS OWNERSHIP

Sole Proprietorship

In a sole proprietorship there is a single owner who assumes the risks and receives the profits (or suffers the losses) from the enterprise. The owner makes the decisions, schedules the work, manages the personnel, determines the credit and collection policies, is responsible for the bookkeeping records, and solves any problems that arise. The owner is able to hire consultants without the approval of others. Starting a new enterprise is faster and easier with a single owner. Changes in the operation are also easier to effect. In fact all business operations are less complicated and there are fewer *regulations* and restrictions that apply to the sole proprietorship as compared to other business types. Clint represents a sole proprietor, Figure 4–1.

Figure 4–1   Many young entrepreneurs start with a livestock enterprise. Clint feeds his ewes; he operates as a sole proprietor. (Courtesy Donald F. Connelly)

One of the drawbacks to sole proprietorship is obtaining the start-up capital and loans. This is due to the single source of original capital (the owner) and that person's ability to provide capital or obtain loans for operation. Another disadvantage is the frequent need to go outside the business for advice or management help.

On the positive side, the sole proprietorship enjoys the simplest tax structure and the fewest restrictions or governmental regulations on business activities.

Partnerships

Partnerships offer an immediate advantage in obtaining capital since two or more persons are the founders and therefore are sources of start-up capital as well as operational funds. Decision making is shared as are all aspects of the business. *Start-up* costs are lower than those in corporations. There are more restrictions and regulations to deal with than in the single owner business but less than in a corporation. In the tax situation, partnerships are easier to handle than the more complex corporation, where both the corporation and the stockholders are taxed on corporate earnings.

Several variations of partnerships exist. Simple partnerships have unlimited liabilities (each partner is fully liable for partnership actions and practices). In a limited partnership, liability is usually limited to the capital invested. Silent partners are those who provide money but usually not the labor or management as contrasted to operating partners who provide mind, money, and muscle to the business.

In forming partnerships, it is important to consider the development of a partnership agreement that spells out the duties and responsibilities of each of the partners and how decisions are to be made and implemented. Such an agreement can prevent many bitter disagreements and misunderstandings over such financial matters as *credit* policy, capital retention, and loans, as well as operating methods.

Among the drawbacks to this type of organization are the unlimited liabilities of each partner, divided management and decision-making authority, and differing outlooks on credit (credit obtained as well as credit granted). A major concern is the match-up of partners who work closely in the business; considerable thought should be devoted to *compatibility*.

The Corporation

The corporation is a type of business organization where the ownership is provided by the purchase of shares. It is a form of multiple ownership. A major advantage of this type of organization is the wide source of capital through sale of stock. The corporate structure is defined as a legal entity empowered to do business as an individual. With the formation of a corporation, there are legal forms to complete and a larger body of restrictions and regulations (as well as taxation) to consider. Don Steele manages a corporate business, Figure 4–2.

The owners of the corporation, the stockholders, are liable only to the extent of their investment in shares. Income to owners is in the form of stock *appreciation,* increased value resulting from the net worth of the company, and by dividends paid on stock holdings. Corporations are the least flexible of the business types, take longer to start, and take longer to effect changes.

The Cooperative

The fourth type of business organization will be discussed only briefly—the cooperative. Cooperatives are organized by a group of producers or consumers so that pooled action will benefit the members of the group. Profits or losses are the responsibility of the participating members. The profits or losses are normally shared in relationship to the amount of business each member does with the cooperative. Ownership is usually maintained by current membership using a revolving fund, sometimes considered shares, that accrue to participating members. For example, a dairy cooperative operating a creamery put one cent per pound of butter fat into a revolving fund certificate. After the start-up loan was repaid and the cooperative was totally member owned, the oldest revolving fund certificates were paid off by new revolving fund certificate monies.

Figure 4–2   Don Steele is president and operating manager for the Copymat corporation.
(Courtesy Don Steele)

Cooperatives trade at current prices although some offer membership discounts instead of the more common patronage dividends.

The Advantages of Different Types of Business Organizations

Figure 4–3 compares the main types of business organizations described above on a number of business characteristics. The form of business you plan to start encompasses all of the ideas expressed in this chapter.

In making the decision about the form of business you wish to establish, you need to consider short-term goals, intermediate-term goals, and long-term goals so that a short-term goal does not lock you into an expensive change in the future. Chapter 3 deals with goal setting in detail. Figure 4–4 is a franchise, a type of business with many aids for the entrepreneur.

Figure 4–5 outlines some of the decisions on business form and gives a reason for such a choice. As you decide on your form of business organization, give thoughtful consideration to these decision points and their effect on your way of managing your business.

ORGANIZATION CHARACTERISTICS	SOLE PROPRIETORSHIP	PARTNERSHIP	CORPORATION
Profit or loss	All	Partners share	Stock share
Ease of starting	Simple	More complex	Most complex
Ease of changes	Simple	Medium	Difficult
Degree of regulation	Lowest	Low	Highest
Taxation	Simplest	Next simplest	Complex
Decision making	Very rapid	Rapid	Slower
Liability	Unlimited	Unlimited	Limited
Start-up cost	Lowest	Low	Highest
Management responsibility	Sole	Shared	Shareholders elect management
Obtaining capital	Most difficult	Less difficult	Sale of shares

Figure 4–3   Comparison of the major types of business organizations. (Courtesy The Three Entrepreneurs)

Figure 4–4   Purchase of a franchise gives an entrepreneur many corporate management aids. (Courtesy The Three Entrepreneurs)

	CHOICES	OWNERSHIP REASONS
Forms	Sole Proprietorship	Complete autonomy over mind, muscle, and money
	Partnership	Divided mind, muscle, and money responsibilities
	Corporation	Liability limited to investment
Types	Product or Production	Preference
	Service	
	Franchise	National recognition
	Non-franchise	Costs of advertising and so on
Methods of Starting	Build from scratch	Complete control
	Buy	Goodwill and name recognition

Figure 4–5 Decision points on business forms. (Courtesy The Three Entrepreneurs)

SUMMARY

In this chapter we have attempted to acquaint you with the three major types of business organization geared to the profit motive.

The sole proprietorship is easily started, easily changed, and easily ended. The owner has full power to make all decisions consistent with laws and regulations. He or she suffers any losses and collects any profit.

The partnership consists of two or more entrepreneurs working together and pooling resources for profit. Management is divided as are profits or losses. The partnership is more complex but much less so than the corporation.

The corporation is a legal entity created to operate as an individual. Corporations are usually owned by a number of persons by means of shares. Corporations are regulated to a higher degree than are sole proprietorships or partnerships.

STUDENT ACTIVITIES

1. Go to your nearest shopping center. List the businesses there and then classify them as to type of ownership as well as their type of activity. You will likely need your instructor's help on some of these.
2. Suggest a partnership agreement for a hypothetical business.
3. Role play the discussion by partners on doubling the size of their roofing business.
4. Organize a hypothetical class corporation. Be sure to meet the legal requirements for corporations.

SELF-EVALUATION

Types of Business Organizations

True-False: On a separate sheet, write T for true statements and F for statements that are false.

1. Businesses are classified by type of organization as well as by type of business activity.
2. A share owner puts his whole property at risk for the business.
3. Silent partners are only liable for their money invested.
4. Sole proprietors are people with silent partners.
5. Limited partnerships find it easier to finance a business.
6. Partnership agreements are important to a smoothly running business.
7. Corporations are more complex than partnerships.
8. Partnerships are more complex than sole proprietorships.
9. In a partnership, each partner can be held liable for actions of the other partner or partners.
10. Divided management responsibility may be either an advantage or a disadvantage in a partnership.

Multiple Choice: On a separate sheet, write the letter representing the best answer for each of the following questions.

11. The corporate form of organization is the best when:
    a. one person wants to control everything.
    b. people want to do business with limited liability.
    c. two people want to start a business.
    d. a group wishes to buy at an advantage.

12. The partnership form of organization is the best when:
    a. one person is the manager.
    b. a group wishes to sell their produce advantageously.
    c. business liability is considered.
    d. two or more persons want to cooperate to form a business venture.

13. A sole proprietorship is the best business organization when:
    a. one person wants to control everything.
    b. liability must be limited.
    c. several people put money up front to start the business.
    d. a group wants to cooperate.

14. Franchised businesses:
    a. are owned by the parent company.
    b. are controlled completely by the parent company.
    c. may be owned by an individual, a partnership, or a corporation.
    d. are large corporations.

15. A corporation is owned and controlled by:
    a. individual shareholders who have equal votes.
    b. share owners who vote as a block.
    c. share owners whose votes equal their shares held.
    d. the board of directors.

CHAPTER 5

Should I Purchase a Business or Build One?

OBJECTIVE

To examine the considerations in buying or building a business.

COMPETENCIES TO BE DEVELOPED

After completing this chapter you will be able to:

1. List six advantages of a purchased business.
2. List six advantages of a business started from scratch.
3. Describe the advantages of a spin-off business.
4. Correctly identify the most important considerations in the purchase of a business.
5. Describe a procedure for evaluating the potential of a business you might consider purchasing.

TERMS TO KNOW

Clientele	Competitor
Consideration	Diversification
Facilities	Family estate
Feed advisor	Feed franchise
Operational	Public relations (PR)
Predecessor	Spin-off business
Suppliers	Ultimate

INTRODUCTION

There are no hard and fast rules for determining when to buy or build a business. The decision about the way to enter a business field most advantageously must be made on a case by case basis. In this chapter we will try to highlight some of the factors for *consideration* in this major business decision.

CASE STUDIES

The Case of the Purchased Competition

Joe operated a well established service station he had taken over from the *family estate* after his father's death. About the same time, Ed purchased the Kelly Brothers Station across the corner. At the advice of their suppliers, they posted identical gas prices and business continued uninterrupted.

After six months, Joe's station seemed almost deserted while Ed's business was increasing. The local Chamber of Commerce secretary decided he would quietly investigate for his own satisfaction. He patronized each station for gasoline and service, and over a period of weeks came to these conclusions:

- Joe gave good service.
- Ed gave good service.
- Joe was aggressive in any area of dispute over fees or costs while Ed took the short side of the ticket if there was a disagreement, just as his *predecessor* had, and customers came back to Ed while they left Joe.

Phil Builds a Business

In another case study, we read about Exeter Feed and Grain being a *spin-off business*. The feed business took the elevator division of Exeter Motors and added a *feed franchise*. While the elevator business was a continuing element of an existing business, the feed business was built from the ground as a major part of the total volume of the new enterprise.

Phil attended a *"feed advisors"* school, performed in-store demonstrations, and on-farm efficiency proofs in order to build customers for his top quality line. In addition, Phil spent a considerable amount of time visiting new and prospective customers during the slack periods at the store, leaving his employees in charge.

What the Cases Tell Us

- You can purchase a certain amount of goodwill, but you must earn it to keep it.
- Over time, business is built or lost by the policies adopted.
- Human relations can make or break a business.

- A purchased business is likely to be active sooner than one started from scratch.
- Two identical businesses are not truly identical. They take on the characteristics of their operators.

STARTING IN BUSINESS

Advantages of Purchasing a Business

The advantages of a purchased business are numerous. Some of these are the established traffic or trade, the name recognition, the cash flow in operation, and the familiarity of the suppliers with the business. The business also has a *public relations (PR)* record, and can be evaluated before purchase. It is an operating facility in place with experienced help, and it may eliminate one *competitor.*

A business in operation has an established trade or traffic pattern regardless of its quantity or quality. One can observe and decide on its potential and improvement possibilities. The location can be evaluated in relationship to all other factors.

The business in operation also has name recognition and purchased businesses often go to great lengths to preserve the value of this in connection with the new name. For example, in a recent banking change the advertising went something like this: "We are now Bank Blank but Sue Jones is still Sue Jones." Another well-known business changed hands almost 20 years ago. Everyone still knows it by its original owner's name but knows the new owner as well. He could change the name with no loss of recognition but sees no need to pay the costs needed to change signs, literature, stationery, and so on.

An operating business has a cash flow in action while a business started from scratch will take time to build up a good cash flow position. Cash flow, as we learned in an earlier chapter, is a demon that curses lots of young businesses. Without cash flow, unlimited capital is needed to stay in business in order to pay the overhead and buy materials. Sales, with the resultant cash flow, are literally the life blood of a business. Many people buy existing businesses as a way of getting started, Figure 5-1.

Suppliers know an existing business. When Phil started Exeter Feed and Grain, his suppliers and grain customers knew two of the partners and Exeter Motors. Realizing that the grain business was a spin-off, they were willing to extend immediate credit for shipments.

Any established business has public relations. They may be favorable or unfavorable depending on given circumstances. In taking over a new business, one should always maintain all positive elements of public relations and try to change the negative aspects by "living right and getting credit for it."

An established business has a track record that can be evaluated before purchase. Obviously a seller will try to accentuate the positive side of the business, but when your accountant goes over the books for you, important facts can be highlighted for further evaluation.

An established business will have *facilities* in place that may or may not be appropriate. They will be operational, however. The cost of any desired changes should be worked into your plans upon purchase. There may be a short- or long-term timeline. The equipment should also be considered in the same light as the facilities.

**Figure 5–1   Entrepreneurship often starts with the purchase of an existing business.
(Courtesy The Three Entrepreneurs)**

An experienced staff can be either a positive or a negative element. They know the business, but they have ingrained methods of operation. To illustrate the point, a few years ago one of the authors took over as acting chairman of his section. His first action was to fire the secretary. After her shock subsided, he asked her if she wanted to work for him. It effectively made the point that old work rules were gone and that somebody new was in charge. A new manager may need to let the public know this also, Figure 5–2.

Purchasing a business as compared to starting a new one sometimes eliminates one competitor. In the case of Exeter Feed and Grain, there was the same grain competition as before but one more competitor in the feed business.

Figure 5–2   **The owner of this business wants the public to know some changes have been made. (Courtesy The Three Entrepreneurs)**

Advantages of a Totally New Business

Among the advantages of starting a totally new business are the opportunity to create your own image, the ability to explore full creative potential, the possibility of starting with totally new facilities and equipment, the opportunity to choose the location, the ability to select suppliers and to choose areas or levels of competition with established businesses.

Sometimes a new business owner has the problem of overcoming a bad image supplied by a previous owner. In a totally new business, the opportunity to create your own image and policies is possible. Policies should always be thought through carefully before they are adopted and changed when obviously inadequate or harmful.

A new business is an opportunity to be fully creative in all matters pertaining to the business. It is not bound by previous history. Within the limits of the business plan and budget, a new business may approach its ideal in the way of facilities and equipment.

The importance of location is carefully detailed in another chapter. Starting a new business has as one of its greatest advantages the ability to choose the most advantageous location. It is said about businesses that there are three important things, "Location, location, and location."

The new owners have the opportunity to choose the supplier even in a purchased business. It will be easier to do so with a new business, however. The quality of the supplies purchased will often have an effect on the quality of the product or service offered.

Another unique advantage of the new business may be the ability to select a particular area of competition or specialization. For example, a fast-food store that opened in West Lafayette,

Indiana specialized in rabbit meat. People quickly made it a popular place as they tried something different for lunch. Figure 5–3 depicts a new business.

Figure 5–3   Ronnie and his father started from scratch in their business and it has grown rapidly. (Courtesy Ronnie Orem)

Considerations in Purchasing a Business

There are some important considerations that should be examined critically before purchasing a business. You will need to explore the same areas as you start a new business, only in a little different way.

The first consideration in a purchase is the reason for the sale. It may be difficult to determine the real reason for the sale. Sometimes this is of *ultimate* importance, and at other times it does not matter at all. As an example, one of the authors knows a restaurant owner who sold a profitable restaurant with limited *clientele*. The location was inadequate but the chef had developed a regular following in the area. The reason for the sale was hard to discern. Careful consideration of all the questions raised here is important. The following paragraphs will detail the major concerns in considering a purchase.

Is the business well managed? If the business is well managed, why is it for sale? If it is poorly managed, will correcting the management errors be possible and profitable? Will you have the needed skills or experience to bring about the needed changes?

Is the business well financed? Financial problems are frequently the cause of business failures. If the business is well financed, how will your resources match up? Will you be able to continue existing credit lines if credit is used to build profitable volume? Will the cash flow provide needed leverage?

Is the business profitable? If the business is profitable, how profitable? Will it remain profitable or even become more profitable under the new management? Will the profits be adequate for your desired goals?

Is the business successful? You ask, if it's profitable, how can it not be successful? If a needed service is being rendered and is the only one available, the business will be profitable under new management. It would not necessarily be successful because if a competitor were to appear it would lose many of its customers.

Is the business seasonal in nature? Although this consideration is not a crucial one, it is important. A highly seasonal business will make it difficult to recruit and retain competent, experienced help. Some of you will likely have summer jobs in seasonal businesses to earn money for college expenses.

Will it be possible to broaden the business and increase business activity at other times of the year? If *diversification* is possible, then permanent employees will be easier to obtain and maintain.

Again, one of the most important considerations is how well the business is located or how location can be turned to a better advantage.

A final consideration may be closely related to the first question. Why is the business for sale? Among the multitude of reasons for the sale of a business are the following:

Death of an owner or partner

Retirement of an owner or partner

Boredom of a smoothly run business

Desire to reduce the hours committed to a successful business

Insufficient finances to make the business profitable enough for the owner

Inadequate profits or actual losses

Whatever the reasons for the sale of the business, one should carefully consider how the correction of the problem will affect the profitability of the business under the new owners.

SUMMARY

Advantages of a Purchased Business

- Established traffic or trade
- Name recognition
- Cash flow in operation
- Suppliers know the business
- PR may be positive (or negative)
- Experience available for study
- Facilities and equipment in place
- Staff may be experienced and in place
- Eliminates one competitor

Advantages of a Totally New Business

- No history to outlive
- No policies to overcome
- Full creative potential
- No existing equipment or facilities to limit startup
- Full choice for equipping the operation
- Opportunity to choose location
- Opportunities to choose suppliers
- Choice of area of competition

Considerations in Purchasing a Business

Is the existing business:
- well managed?
- well financed?
- successful and profitable?
- seasonal?
- well located?

SELF-EVALUATION

1. What are six advantages of a purchased business?
2. What are six advantages of a business started from scratch?
3. What are the advantages in a spin-off business?
4. What are the important questions to ask when starting or buying a business?

STUDENT ACTIVITIES

1. Look up the businesses for sale in the classified section of your newspaper and identify types that are for sale.
2. Interview an owner who has just purchased a business and ask why the business was for sale.
3. Have the students bid on the purchase of an imaginary insurance business created by file folders giving the name of the customer, his insurance policies, and their premiums.
4. Contact a local realtor and have him or her come to class to tell how to locate a business that might be purchased.

CHAPTER 6

What Are My Three Business Resources?

OBJECTIVE

To develop an understanding of your business resources.

COMPETENCIES TO BE DEVELOPED

After completing this chapter you will be able to:

1. Describe the three business resources of mind, muscle, and money.
2. Complete a personal balance sheet.
3. Compute net worth given total assets and liabilities.
4. Explain the concept of net worth.
5. Identify your assets in terms of mind, muscle, and money.

TERMS TO KNOW

Consume

Deflation

Field representative

Inflation

Relocation

Contact people

Direct writer

Independent

Interest payment

Transaction

MATERIALS NEEDED

List of your assets and liabilities.

INTRODUCTION

The objective of life is to manage yourself in order to use what you *now* have (your mind, muscle, and money resources) to get what you *now* want most. Isn't your goal to get as much satisfaction as is possible from the use of your present resources?

Some persons do a better job of managing the use of their resources than do others. They get more satisfaction from the use of their mind, money, and muscle. We say these persons are better managers of their lives. You probably want to be a better manager. You want to get as much satisfaction as possible from your work as well as your leisure activities. By participating in this program, you will learn to improve your management skill. You will learn to make better choices.

CASE STUDIES

The Case of a Goal Achieved

Don and Janet had been pursuing the goal of owning an *independent* insurance agency. Don had spent 5½ years as a *direct writer* for a major insurance company. Through their many contacts they had indicated this interest. A *field representative* of another major insurance company knew of an older agent whose health was failing and put buyer and seller in touch with each other.

The field representative had to approve the new agency as representing his company. Don's experience and acquaintance with the field representative assured the needed approval. The purchase of the general insurance agency also required *relocation*. The new location was in an area that suited both Don and Janet and thus was not a problem.

The Case of an Existing Resource

Howard and Barbara bought some acreage at the location of their new job. The next May, people showed up to pick strawberries. The former owner had grown an acre of strawberries and sold pick-your-own berries. Since the field was bigger than Howard and Barbara could use, they helped their children organize a strawberry enterprise. This grew into the Dan-D Acres Corporation.

What the Cases Tell Us

- There are many *contact people* who know a business well enough to help locate opportunities for you.
- Your current resources provide a basis for advancement.
- Experience in a field is a valuable resource.
- Experience will aid in financing a new enterprise.
- Experience is a large part of the mind resource.

- Savings are a valuable part of money resources.
- Good health is needed for the muscle resource.

IDENTIFYING YOUR RESOURCES

What Do You Have?

Who are you? What resources and skills do you now possess? If you want to use what you have to get what you want, you need to determine your present position. You can do this by making an inventory of your present situation in terms of your mind, muscle, and money resources. Clint demonstrates one of his resources, Figure 6–1.

Figure 6–1 **Mind, muscle, and money are the entrepreneurial resources. Most young entrepreneurs have more muscle than money in the beginning. (Courtesy Donald F. Connelly)**

What is Your Financial Position?   People measure their financial position by listing all their property with an estimate of its present market value. They also list the amount of their financial debts. People call this report a balance sheet. Businesses use a balance sheet and lenders require one as a reporting device or as a statement of financial position for obtaining capital funds. You can learn about a balance sheet by completing one for your own use. Figure 6–2 provides an outline for your simplified balance sheet. (Obtain a blank copy from your instructor instead of using the copy in the text.)

MY BALANCE SHEET AS OF TODAY'S DATE

Assets (something I own)		Debts (something I owe)	
Cash		Due to be repaid in the next year	
in my pocket	_____	Accounts payable	
in my savings account	_____	to my relatives	_____
in my checking account	_____	to others	_____
		notes or mortgage	
Other Assets		payments	_____
Clothes, bike, etc.	_____		
Auto	_____	Due to be repaid	
Business inventory		after 12 months	
goods in process	_____	to relatives	_____
Farm equipment	_____	to others	_____
Other business		notes or mortgages	_____
equipment	_____		
	_____		_____
Total Assets	$ _____	Total Debts	$ _____

Figure 6–2   A simplified balance sheet. (Courtesy The Three Entrepreneurs)

To find your net worth, subtract the amount of your debts from the value of your assets. Your net worth thus is your financial worth. *Note:* The formula is **Net Worth = Assets − Liabilities (Debts).**

How Do You Calculate Worth?   How did you obtain your financial net worth? As you will study in more detail in another chapter, you accumulate financial worth in three ways. First, you can receive cash or other property by gifts or inheritances from others. A few persons receive large amounts of money or other property in this way. You might consider yourself fortunate if you are or will become such a person. Second, once you own property, you sometimes realize an increase in its value because of *inflation* (or a decrease in value due to *deflation*). This is recognized as a change in a valuation account in your net worth. Figure 6–3 is a complete balance sheet showing net worth.

Finally, you can use your mind, muscle, or money resources in exchange for money. Figure 6–4 illustrates the use of non-money resources.

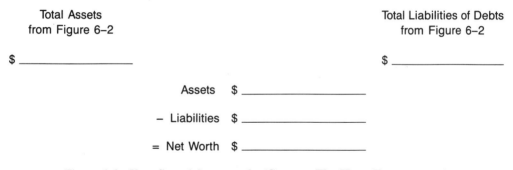

Total Assets
from Figure 6–2

$ _____

Total Liabilities of Debts
from Figure 6–2

$ _____

Assets $ _____

– Liabilities $ _____

= Net Worth $ _____

Figure 6–3 Your financial net worth. (Courtesy The Three Entrepreneurs)

Figure 6–4 DAN-D-Acres started with a resource found when the Doster family relocated to a small acreage with a berry patch. (Courtesy DAN-D-Acres)

1. You can work your mind and/or muscle in exchange for a salary or wage.
2. If you already have some money, you can put it into a savings account or loan it to someone else in exchange for a cash *interest payment.*
3. You can also try to increase your financial net worth by creating your own business. You can combine the use of your mind, muscle, and money resources to create a new good or service. While your expectations are not always fully realized, you expect to make a profit in this business. That is, you expect to sell your newly created good or

service for more than you spent to make it. Any excess money is called profit or earnings. It is a return to you for the use of your resources, a newly created earned net worth.

No doubt you will decide to *consume* some of this net worth. You may decide to use some of your remaining net worth in your own business or lend some of your money to others in exchange for interest payments.

One way to measure the financial success of any decision you implement is to calculate the amount of earned net worth you have realized from that venture. Certainly, a significant measure of your performance in your entrepreneurial activity is the change in your net worth because of it.

How Often Can You Measure Net Worth?   Net worth can be measured by following three steps. First, place a dollar market value (using third party appraisal discussed in a later chapter) on all property that you own. Second, subtract the total amount of your financial debts. Third, recognize the difference in value as your current net worth.

You can calculate your net worth several times each day. You might consider doing this after each financial *transaction* you make. What will you discover? If you make appropriate calculations, you will find that some of your transactions cause your net worth to decrease, others cause no change in net worth, and still others cause an increase in net worth.

How is Net Worth Affected by Different Types of Transactions?   Net worth effects will be discussed in more detail later. For now, consider the following illustrations.

Your financial net worth is decreased when you buy *and consume* a sandwich or go to a movie. You needed the food to satisfy your appetite and wanted the movie for entertainment. The gain you realized from these activities is offset somewhat by the financial pain you now recognize in a lower net worth.

When you choose to consume net worth, you no longer have the same amount of financial resources available to use in the future. You have less money to save or lend to others in exchange for an interest payment or to use in your own business.

The decisions you make about consumption versus savings are significant. When you save money, you have more financial resources to be used along with your mind and muscle resources. When you have more of your own money, your business is affected in two ways. First, you are able to borrow more money because lenders feel more secure about your ability to repay them. Second, for the same size business, you will need to borrow less, and thus your interest expense will be lower.

Moderation is a key to success. You will probably choose to consume some of the money you earn. It is a way to realize many of the satisfactions in life. Remember, however, the likely result of consuming much of your net earnings is that you will never own a very big business.

Some financial transactions do not affect your financial net worth. When you buy equipment or fertilizer, you increase your values in one type of asset and decrease your values in another. When you repay a loan, you increase an asset and decrease a debt. None of these transactions affects your net worth.

Some transactions do affect your net worth. When you receive wages, rent, or interest for the use of your resources over a period of time, you increase your cash and therefore your net worth. When you sell a good or service at a profit, your net worth increases. (In addition, your cash increases and your goods in process inventory decrease.) Profit (earnings) is the main source of net worth increase in most businesses. It is perhaps the single best measure of your business performance over a period of time.

You probably know families with similar earnings that vary greatly in their net worth because of the management of their assets. It is also true that it is hard to start or maintain a business if a person has no financial net worth. A lender is much more likely to approve a loan to a person if that person can show a past record of earned net worth increase.

Finally, a person is likely to proceed into a new business venture with more confidence in the ability to perform successfully if a past record of earned net worth increase has already been developed. Past performance is the best measure for estimating future performance!

What is Your Mind Position?   You answer questions on classroom tests quite often. Each of these tests is intended to measure what you *now know* about a particular subject. Your scores on formal tests are one measure of your mind skills on the date you take the tests.

Prospective employers often ask to see your school records, including your attendance record. They want to know what you have done in the recent past. Employers may also ask you to take one of their tests to see how well you do compared with other people who are applying for a job. Previous on-the-job performance involving decision making, supervising, or quality of work may also indicate your level of mind skills.

Perhaps the opportunity to improve your mind skills is the main reason you are involved in this entrepreneurial activity. Good practice improves one's skills and certainly you and I believe we can learn to improve our performance. Part of the learning process includes becoming aware of what is possible. Another part is actual practice in various skills.

What is Your Muscle Position?   How fast can you run a race? How many push-ups could you do last month? Today? You can think of many fun, work-related tests to measure your muscle skills. For now, recognize that many young people primarily have muscle resources. These resources can be used in one's own business. On the other hand, a person can work for a wage paid by someone else.

If you have mostly muscle resources to offer, use them wisely and carefully. Most people don't like to do muscle work, so you may find significant opportunities to perform jobs that others don't want to do. Even when you find that you are not able to devote your time to mind and/or money endeavors, remember your muscles. A healthy body is perhaps the single most envied resource.

How Can You Use What You Have?   Look again at your inventory of your mind, muscle, and money resources. You measured them as of today's date, *a point in time*. However, you use resources only during a *period of time*. Money is loaned for a day, a month, a year, and so on. You earn a wage or salary for similar time periods. The earned net worth you now have is the accumulated value from all resources you used in previous periods. You transformed the use of your mind, muscle, or money resources into goods or services that you sold at a profit. Figure 6–5 shows additional resources available to us.

Mind and muscle skills deteriorate if not used. Maybe, at one time, you could do 20 push-ups. If you haven't done any recently, you may find you can now do only five. Recognize that you must use each of your resources or you lose the value you might have realized from their use.

Figure 6–5   Family and friends are valuable resources. (Courtesy DAN-D-Acres)

How Much Time Do You Have Available?   Everyone starts with the same amount of time. You have 24 hours in each of your future days. Remember, you checked this in Chapter 3. How do you want to use your resources? Likely, you will decide to continue to use much of your mind and muscle time for learning and leisure activities and much of your money resources will continue to be used in their present form.

How much of your resources could you commit to an entrepreneurial activity? Take a calendar and think back over how you used each day last year. Think about what other things you plan to do for each of the days in the coming year. How much mind and muscle time each day, each week, and each month could you make available?

Remember what we said in Chapter 3: *Keep your priorities in mind as you plan each day, week, or year.* These priorities may change over time.

SELF-EVALUATION

Resources Test

Please give brief but complete answers to the following questions:

1. What are a person's resources?
2. List the elements necessary for a balance sheet.
3. Make a pie chart representing a 24-hour school day. Note time in which you can choose your activities.

STUDENT ACTIVITIES

1. Find company balance sheets to bring to class for discussion.
2. Complete a balance sheet and net worth statement of this student: The student owns a 10 speed bike worth $175, a car worth $1,250, clothes worth $750, and a stereo worth $115. He owes his father $1,050 for the car and his brother $10.

CHAPTER 7

What Is My Competition?

OBJECTIVE

To recognize the nature of competition and its importance to a new business.

COMPETENCIES TO BE DEVELOPED

After completing this chapter you will be able to:
1. Identify the results of consumption vs investment decisions.
2. Identify examples of elastic or inelastic demand.
3. Identify products that can be substituted for each other.
4. Describe the effects of supply on prices.
5. Identify market-clearing price.
6. Identify competing businesses.

TERMS TO KNOW

Alternative price	Bankrupt
Competitive markets	Competitor
High cost producers	Hypothetical
Investment	Low cost producers
Market economy	Outside changes
Recipe	Scarce or scarcity
Supply shifters	Technology

MATERIALS NEEDED

A mall map showing businesses in a nearby mall.

INTRODUCTION

What's my competition? That's an important question. You may consider it every day for the rest of your life. Competition can be thought of in several ways. One way is to look at it as the process you go through in deciding how to use your resources. How much will you consume now? How much will you use to produce goods and services now in order to increase your own consumption in the future? A second way to consider competition has to do with your choice of specific consumption items. Finally, as a business person, you attempt to evaluate the performance of others, your competitors, as you consider what to produce, how to produce it, how much to produce, how to market it, and how high to price it.

In starting a business it is very important to know what and who your competitors are. Careful market research should be performed in order to develop a realistic cash flow projection. Figure 7–1 illustrates competition in fast foods.

Figure 7–1   A high level of competition in the fast-food business is evident at this location. (Courtesy The Three Entrepreneurs)

CASE STUDIES

The Case of the New Seafood Shop

Hal had worked with his father and other local fishermen in a small city along the coast. As they marketed their catches daily, their return was only a fraction of the price received by the retailer. Hal thought that he saw an opportunity to get out of the long, cold, wet hours in the boat and improve his lot as well as that of the other fishermen. After budgeting carefully, he started a seafood sales shop in his city. Hal was just starting to do well when the major supermarket chain decided to expand its "Fresh from the Sea" department. Hal learned about competition in a hurry.

The Case of the Sideline Competitor

Exeter Feed and Grain had not been in business long before Phil learned that a brand of feed not sold in any Exeter store would appear on many of the farms he serviced. Discreet inquiry found that a local trucker had managed to buy the feed wholesale at a terminal market and then sell it at a reduced price in order to have a return load from the market city. Although it was a lower quality product than that which Exeter Grain sold, it did cut into sales.

Phil quit using that trucker to haul grain to the elevator. He told the trucker he could not pay an unfair *competitor* when there were other truckers available to haul the grain who were not competing in the feed business. Figure 7–2 illustrates another competitive business.

What the Cases Tell Us

- Market research is essential.
- Competition may be based on convenience, quality, service, price, or a combination of these and other factors.
- Knowledge of one part of an industry does not always apply to another part of the industry.
- When a business is perceived as profitable, others will enter the market.
- Some consumers are only price conscious; others recognize both price and quality.

WHAT CAUSES COMPETITION?

Personal Investment/Consumption Decisions

Persons make different value judgments. Persons with about the same resources may differ greatly in their *investment* versus consumption decisions. One person may choose to buy a new top-of-the-line car. Another may choose to increase a savings account. A third person may invest in a new piece of equipment for business. Presumably, they are equally satisfied with their decisions, even though the likely consequences are quite different.

You face this type of decision daily. Occasionally, you will make a major consumption versus investment decision that will have a major impact on your life. What decisions regarding investment versus consumption have your parents, or someone else you know well, made recently? What did they decide? What factors did they consider? Would you have chosen the same alternative?

Remember, there are not necessarily right or wrong investment versus consumption decisions. However, the decisions often have quite different consequences.

**Figure 7–2   Pick-your-own strawberries is a competitive enterprise. Young owners
profit well from this entrepreneurship. (Courtesy DAN-D-Acres)**

Demand

The concept of demand describes the behavior of people as they fulfill their desires for
consumer and producer goods and services. The concept can be observed in the market place.
Demand refers to the quantities of a product that people are willing and able to buy at alternative
prices, while holding everything else constant. People buy more of a product at lower prices
and less at higher prices. For example, there are more Chevrolets, Fords, and Plymouths on
the roads than Cadillacs, Lincolns, and Imperials.

Because we all want more things than we can buy, we must choose how to use what we
have to get what we most want. We consider the satisfactions we can expect from consuming
various amounts of our resources. In our *market economy,* we study the trade-off between our

satisfaction and the market price of the goods or services. Entrepreneurs often get more satisfaction from investing in their business than from a consumption decision.

Some goods and services are extremely important to us. If we're diabetic, we will pay almost anything for our daily dose of insulin. We won't pay much for any extra insulin, however. We have what is called an inelastic demand for the item. We will pay any price if we need the product and can get it, but we don't want much extra even if the price were to drop considerably. In other words, inelastic demand means that the same amount is needed regardless of price.

Most farm grain crops have inelastic demands. There are few good substitute products for grain, either for human food or for livestock feed. People want to eat. On the other hand, they won't pay much for extra food when they are already full. How many ice cream cones will you eat after you've already had seconds of Thanksgiving turkey?

A different demand situation occurs with different types of foods. A small change in price between hamburgers and hot dogs may cause a big change in quantity of hamburgers eaten or demanded. Because there are many substitutes for hamburgers, the quantity demanded changes greatly with a small change in price. This product is said to have an elastic demand. In an elastic demand, a small change in price can mean a large change in sales. Persons in similar restaurant businesses must be careful not to price themselves out of the market.

Automobiles also have an elastic demand. If the price of a particular brand changes a little bit, the quantity demanded changes a lot. Autos represent a large purchase to most people. They shop carefully for the best deal in making large purchases.

What kind of product or service will you have? Does it have an inelastic or elastic demand? Will the total number of sales, both yours and those of your competitors, change much if you offer to sell it at a slightly lower price or with slightly better quality? Young entrepreneurs find an area for successful competition in Figure 7-3.

Although the demand for corn is relatively inelastic, the demand for a specific farmer's corn is extremely elastic. One farmer's corn is almost exactly like every other farmer's corn. If one farmer were to ask a slightly higher price, he or she would sell no corn. Corn grown on other farms is such a good substitute that no one will buy the higher priced crop. Alternatively, if a farmer lowered the price a little, he or she would sell the entire crop first.

Many people purchase corn, and it differs little except for location. Therefore, the decisions individual farmers make have little effect on price. Individual farmers have almost no effect on corn prices, therefore we say they are price takers not price setters. Their main pricing decision is based on *where* to take the price.

What is your market situation? For example, how much location advantage do you have? Are present suppliers of your market now transporting the product long distances to your potential customers? Are your potential customers now traveling long distances to obtain your type of good or service? If your answer is "yes," you may have a significant production and sales opportunity.

Earlier, we said that demand refers to the quantities of a product people are willing and able to buy at different prices while holding everything else constant. But everything doesn't stay constant. Preferences of people change because of advertising among other reasons. Prices of substitutes change. The number of people in a market, their incomes, and their expectations change. For any of these reasons, a different quantity may be demanded at each of the *alternative prices*. Watch for these changes. You will want to adjust your own business accordingly.

Figure 7–3 Lawn and landscape businesses are areas in which young entrepreneurs can compete successfully. (Courtesy Donald F. Connelly)

Supply

The concept of supply can be defined as the quantities of a product that producers are willing and able to offer for sale at various prices, other things remaining unchanged. The quantities supplied will increase when prices for the product are higher. Producers are then willing and able to produce more of the product. They can pay more for the goods and services *they need* for production.

The quantity of products available at any price is directly related to the producers' costs. They must buy what they need from others. They cannot continue to buy for long unless they sell their product at a profit.

Some producers have lower costs than others. First, they may already own important resources, like good farmland that is used to grow corn. Second, they may use their needed resources more efficiently. That is, they may do a better job of selecting a corn-growing *recipe* or of actually growing the corn.

Will you be a *low cost producer?* Having lower costs than your competitors is extremely important if you expect to remain in business. How can you be competitive with others? What can you do better? What can you do at a lower production cost?

In the section on demand, two concepts were discussed: demand as the quantity of goods desired at various prices and the shift in demand brought about by *outside changes.*

In this section on supply, we have described supply as the quantity of goods marketed at various prices. We now recognize that a shift in supply can occur because of outside changes. These can include an improvement in *technology,* a change in resource prices, or a change in the price of substitutes. What *supply shifters* have occurred in your type of business recently? Suppose the price of your product dropped. Which of your competitors would quit production first?

Market and Prices

Production resources are *scarce.* There is an active market for the use of almost everyone's mind, muscle, and money resources. In our market economy, prices change to reflect the effective demand people have for the use of each others' resources. When prices change, the market-clearing position between the quantity demanded and the quantity supplied for any product is affected.

In *competitive markets,* supply and demand represent the decisions of many persons who are willing and able to buy or to sell goods and services. A market-clearing price occurs when the quantity demanded just equals the quantity supplied. This is illustrated in Figure 7–4, where *hypothetical* data are provided for a hypothetical product. The data in the figure are presented in tabular form as supply and demand schedules. Columns 1 and 2 of the figure constitute the supply schedule, while columns 2 and 3 constitute the demand schedule. Column 1 gives the quantities producers would supply at the price in column 2, while the amount consumers will buy at the price shown in column 2 is given in column 3.

QUANTITY SUPPLIED BY PRODUCERS (millions of units)	PRICE ($ per unit)	QUANTITY DEMANDED BY CONSUMERS (millions of units)
70	$7.00	10
60	6.00	20
50	5.00	30
40	4.00	40
30	3.00	50
20	2.00	60
10	1.00	70

**Figure 7–4 A hypothetical supply and demand schedule for a hypothetical product.
(Courtesy The Three Entrepreneurs)**

Figure 7–4 shows that the market-clearing price for this hypothetical product is $4.00 per unit. At any price above this, the quantity exceeds the demand, and competition among sellers will cause the price to fall to $4.00. Only at the $4.00 per unit price will the market clear.

The schedules presented in Figure 7–4 can also be used to show that 40 million units is a market-clearing quantity. Only at this quantity will the market clear, with the price sellers are willing to accept and the price buyers are willing to pay equal to each other.

The market-clearing price of $4.00 and the market-clearing quantity of 40 million units shown in Figure 7–4 will persist only as long as other things remain constant. If demand is increased, prices first rise to ration out the available supply. Producers earn extra profits until they or others increase production. Then, prices fall to a new market-clearing level. If demand is decreased, prices drop to clear out the available supply. Some *high cost producers* become financially discouraged. They then either switch the use of their resources to another business or they lose their financial resources and become *bankrupt*.

Thus, the market works to allocate the use of our mind, muscle, and money resources. The market has no feelings of sorrow for losers or happiness for winners. The people involved do have feelings and are sorry for the losers. Yet we generally recognize that our creation, the market economy, is the best way we know for allocating our scarce resources to best satisfy our many wants. Further, by understanding how our market system works, we have a better chance of using it to our advantage. In this class, we can increase our market understanding skills. We will use these skills again and again.

SELF-EVALUATION

Competition Test

1. Indicate which of the following items are consumption and which are investment decisions:
   a. purchase of a food item
   b. purchase of a mower
   c. purchase of a bond
   d. purchase of a ticket
   e. deposit in savings
   f. purchase of gasoline

2. Indicate which of the following are elastic demand items, generally speaking:
   a. medicine
   b. shoes
   c. wheat
   d. meat

3. Name three products that illustrate the principle of substitution.

4. What is meant by a market-clearing price?

5. Using your mall map indicate the businesses who compete directly as well as those who compete to a lesser degree.

6. Describe how supply and demand affect prices.

Multiple Choice: On a separate sheet, write the letter representing the best answer to the question.

7. The strongest competition to your hamburger stand would be
   a. a hot dog stand.
   b. a restaurant.
   c. a grocery store.
   d. a meat-packing plant.

8. People with approximately the same money resources will make purchasing decisions
   a. in about the same manner.
   b. in the same quality range.
   c. in greatly different manners.
   d. in slightly different manners.

9. Most grain crops have inelastic demands because
   a. one person's is like another's.
   b. shipping costs add to the price.
   c. location of the supply is important.
   d. there are no good substitutes for grain.

10. Hamburgers could be said to have an elastic demand because
    a. one can eat more than one at a time.
    b. there are other foods that can be substituted for them.
    c. they furnish a quick meal when you are hungry.
    d. there are lots of hamburger shops.

11. Your competitive position may be favorable because
    a. there are 10 other suppliers in the city.
    b. your location is best considering freight or other charges outside competitors have to pay.
    c. your price is higher.
    d. your quality is equal to the competition.

12. The quantities offered for sale will increase with
    a. new producers in the market.
    b. fewer producers in the market.
    c. lowered cost of production.
    d. increased prices.

True-False: On a separate sheet, write T for true statements and F for false statements.

13. There are no right or wrong investment versus consumption decisions.

14. Most food stores have an inelastic demand.

15. Supplies will expand to meet the demand at a profitable price.

16. One area for competition is in quality.

17. Lowered production costs will increase sales.

18. Lowered production costs may improve market share.

19. Some producers may have lower production costs because they may own more of the resources needed.

20. The market-clearing concept relates to the point where supply exceeds demand.

STUDENT ACTIVITIES

1. Decide what business you would enter in your community if you had $300,000 to invest and if competition were to be the deciding factor. Compare your answers to those of the rest of the class.

2. How many grocery stores are there in a 25-mile radius? Gasoline stations? Car washes? Fast-food stores? How has competition affected these businesses in your area? Discuss this as a class activity.

CHAPTER 8

How Do I Develop My Market Niche?

OBJECTIVE

To explore the concept of a market niche.

COMPETENCIES TO BE DEVELOPED

After completing this chapter you will be able to:
1. Identify market niches.
2. Describe a technique for doing market research.
3. Identify areas of market saturation.
4. Identify ways a business can expand its market niche.

TERMS TO KNOW

Caters   Demographic
Market niche   Pertinent
Polled   Precision
Research

MATERIALS NEEDED

The map of a mall used in a previous chapter.

INTRODUCTION

A *market niche* can be defined as a situation or activity specially suited to the abilities or character of a business. There are countless examples of successful entrepreneurs who found the right niche in which to market their products. But for every entrepreneur who finds the right niche, there are many others who do not. Finding a market niche is not a hit-or-miss proposition. It is the result of careful analysis.

A market niche is a defined segment of the marketplace. Consider some of the businesses with which you are familiar. The McDonald's restaurant chain, for instance, has found its niche in the fast-food market. While there is a huge market for restaurants which offer leisurely candlelit meals in plush settings, McDonald's is not interested in this type of customer. Rather, the fast-food restaurant *caters* to the consumer who wants a tasty but inexpensive and quick meal. Through *research,* McDonald's recognized the consumer's priorities of speed and convenience, and made the drive-up window a part of all their restaurants. They focused on their specialized niche within the total market and developed it.

Think of the total market as a pie-shaped diagram like the one illustrated in Figure 8–1. Your niche in the market is a small, but very specialized, "piece of the pie." In this chapter, we will give you some ideas that can help you find and develop your market niche.

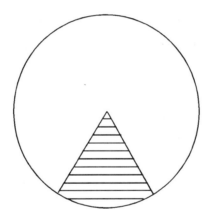

Figure 8–1   Your "piece of the pie." (Courtesy The Three Entrepreneurs)

CASE STUDY

The Case of the Right Product in the Wrong Place

Nancy Miller's family had been collecting small heirlooms for generations. Her parents' attic was filled with old dishes, lamps, doilies, and other odds and ends from past eras. When she inherited her grandmother's estate, Nancy had more heirlooms than she could use. One day, as Nancy was polishing an old silver tea service, she made the decision to open an antique shop. With some research, she could catalog and price the items, and she had enough money saved to lease an inexpensive shop in her neighborhood.

Within six weeks, Nancy hung her shingle over her new shop. She had created an exclusive boutique image and had advertised her business in the *Antique Digest* as well as in the local paper. She settled into her shop, anticipating hoards of well-dressed women leisurely browsing through her aisles of small antiques.

For the most part, Nancy had planned well. She had done her pricing research and knew the market value of her antiques. She had signed a six-month lease with a nearby shopping mall. The atmosphere in her shop was cozy and comfortable, and the items she hoped to sell gleamed in their display cases. She planned to go to sales and auctions and to hire part-time help when she was ready to restock her merchandise. But Nancy's dreams didn't materialize. She sat in her shop day after day, waiting on only four or five customers. Nancy had failed to recognize one of the most important aspects of entrepreneurship. She was trying to market her product in the wrong location.

In order to lease space as inexpensively as possible, Nancy had located her shop in a small mall which included a shoe store, a small grocery, and a drugstore. Customers came to the mall for necessities, but few spent time browsing. While the location of the shop was convenient for Nancy, it was not convenient for people who were interested in shopping for antiques.

From the beginning, Nancy's costs outweighed her earnings, and soon her rent had eaten a large portion of her savings. She saw that her business would fail quickly without some drastic change.

About 15 miles away, there was a small community which had become an exclusive shopping area, drawing an upper-class clientele from a large part of the state. It included several shops similar to Nancy's, shops that specialized in items not normally found in shopping malls. In fact, the marketing concept was exactly the opposite of most big malls today. The shops were located in older restored homes that lined the Early American brick streets of the community. Figure 8–2 illustrates a very specialized business.

Although it required a greater financial commitment from Nancy, she moved her antiques from her local shopping mall to the small, quaintly restored village. In less than six months, she hired two full-time employees to staff her growing business.

What the Case Study Tells Us

- Nancy discovered one of the most important tools of business planning: market research. Her product was not in enough demand to attract customers. She had to take her antiques to an area frequented by potential customers.
- While her new location was not as convenient for Nancy, it placed her antiques in a marketing area that drew the type of customer Nancy needed.
- Nancy's research helped her price her merchandise competitively. She should have carried the research further to check the preferences of potential customers. For example, she might have spoken with other people who were interested in antiques. She could have asked such questions as: "How far do you travel to purchase antiques?" and "If you were searching for a special antique, would you travel out of your way, or would you limit your shopping to a location where a variety of such specialty shops is available?"

Figure 8–2   Just as an antique shop is a specialized market niche this shop illustrates a well-defined market niche. We might call it "the itch niche." (Courtesy N.J. Plantenge)

WHAT IS MARKET RESEARCH?

To be successful, every business manager must find the answers to important questions such as these:

- Who are my potential customers?
- What kind of people are they, and what are their habits, opinions, and preferences?
- Where do they live?
- What do they want?
- What *can* they buy?
- What *will* they buy?
- Do I offer the items where and when they want them?
- Will they pay my prices?
- Will my advertising attract customers?
- Does my business appeal to my customers?
- Who is my competition and what is their sales volume?
- Can I compete?
- What long-term trends will affect my business?
- How do I research to expand my product or service lines?

The only intelligent way to either start a new business or to further develop an existing business is to find the answers to the questions listed above. Market research is, very simply, an organized search for the answers to these questions. The American Marketing Association states it more specifically as ''the systematic gathering, recording, and analyzing of data about problems relating to the marketing of goods and services.''

SHOULD I RESEARCH MY MARKET?

You research the market for one of two reasons. The first is to analyze the opportunities for starting a successful new business. The second is to find ways to improve an existing business. Your goal is to either find or develop your own niche in the market. To do so, you need to consider these three steps.

Define the Problem

Exactly what information do you need? If you are starting up a new business, you need to know if there is potential for success. If you are trying to develop or expand an existing business, you need to know what would attract new customers or what would make current customers become better customers. If you are trying to correct a problem, you need to look at the symptoms of the problem (declining sales or profits) and seek out the cause of the problem (new competition? changing customer profile?).

Decide if Research is Needed

Perhaps the answers are already available. There are several existing sources for market information. The research librarian at your local library can help you find *pertinent* census information as well as a list of trade journals and relevant publications. The Small Business Administration prints statistics and maps for national market analysis and is an excellent source for nationwide information. Your local Chamber of Commerce may have compiled consumer data which pertains to your community. You don't want to waste time and money putting together *demographic* data that is already available and free. Simple market research can be done from home, Figure 8–3.

Decide if the Research is Justified

You have defined your problem and have decided that some research is needed. Now you need to weigh the cost of the research you decide to do against the benefits you hope to derive from it. How much will a wrong decision cost compared to the potential profitability of a correct decision? If further research is justified, you are ready to tackle the research itself.

HOW DO I RESEARCH MY MARKET?

There are several specific steps that you can take which will help you research your market area.

Figure 8–3 Simple market research is as close as your telephone. (Courtesy Donald F. Connelly)

Define Your Objectives

The first step is to decide exactly what information you need and want. Do you need information about new customers or your current customers? Do you want to try a new product or a new service? Do you want to improve sales or increase profits?

Collect the Data

Once you know what information you need, you are ready to obtain it. Marketing experts tell us that the best source of information is the customers. The next best source is a similar business, even a competitor. There are several methods for gathering information. Names, addresses, and telephone numbers are printed on checks. These can give you geographic information about customers. Sales receipts with names and addresses let you cross-reference your customers' locations with the products that they buy. (This information can help you decide where to emphasize your advertising.) A few afternoons spent people-watching in your competitors' stores can give you valuable information about their buying habits. Many businesses keep a file of complaints, which can sometimes pinpoint a problem. The following are some additional methods commonly used to gather data about consumers.

Mail Surveys.   Questionnaires sent through the mail are one means of addressing potential customers directly. One of the drawbacks to this method is the low rate of response. Questionnaires are frequently pitched into wastebaskets, especially when there is no response incentive (such as the promise of a discount coupon for merchandise). The advantage to questionnaires, however, is that they can be made as detailed and specific as you like. It is important that questions be direct and easy to answer. A simple multiple choice question is more likely to be answered than is a lengthy fill-in-the-blank question.

Telephone Interviews.   Another method of gathering information is to speak directly with potential customers. Be especially considerate if you select this method. Some people resent the intrusion into their personal sphere. Express your appreciation for their cooperation.

Face-to-Face Interviews.   If you are considering making changes to an existing business, talk with your customers. They know best what it is about your business that attracts them, and what it is that they would like changed. If you are starting a new business and have no customers of your own, you can talk with people in the vicinity of your prospective shop. Speak with those customers who shop at businesses similar to the one you plan to start.

Employee Interviews.   If you plan to make changes in an existing business, talk with your employees. They are your closest contact with your customers. They overhear both positive and negative comments and therefore may have some helpful suggestions.

In any interview, whether it is with an employee, a customer, or a competitor, it is crucial that you be neat, courteous, and professional. Plan your questions carefully and be direct. If you ask, "Do you think this sounds like a good idea?", you are likely to get an encouraging answer (one that is nice, but not very helpful.) A question such as "Would you drive X number of miles and pay X dollars in order to buy this product?" is direct and personal, and will elicit a more honest response.

It is also a good idea to adjust the scope of your research to the market you want to penetrate. If you have a product you hope to market nationwide one day, then by all means, use whatever nationwide data you can obtain. On the other hand, if you are offering a service limited to your immediate neighborhood, then limit your research to the immediate area.

Organize and Interpret Data

Once you have gathered the information you need, it's time to put it all together. If you used an interview method of gathering data, you could make a master sheet to tally all of the responses to your questions, and in this way study the opinions of the majority of the people. What percentage of the people you *polled* would shop at your new store? How many of your current customers would make use of extended hours? Have your competitors attempted changes similar to the one you are contemplating, and if so, were the changes profitable? Why or why not? The responses you need will depend on the problems you are attempting to solve.

Make Your Decision

When you have done a good job of gathering and interpreting your data, your decision will be simpler and more reliable than a guess based only on your personal feelings about the solution to your problem.

Watch the Results

You have defined your problem, gathered and evaluated pertinent information regarding a solution to the problem, and made a decision about how to deal with it. Examine the results of the changes you make. Did you solve the problem in the best possible way? Was the outcome what you expected? Would you decide differently if you could do it again? Learn from your mistakes. Business management involves making decisions every day, and experience can become one of your most valuable assets.

PLAIN OR FANCY RESEARCH

The word "research" makes us think of scientific *precision* and carefully ordered data. Research does not have to be scientific in order to be effective. Here are some examples.

A restaurant manager has introduced some new items on the menu. He has his buspeople keep a close eye on uneaten food as they clear tables to determine whether or not the new foods are being eaten.

The owner of a new hardware store offers her customers all the roasted peanuts they can eat as they browse through her store on opening day. At the end of the day, trails of peanut shells throughout the store tell her which displays drew the most attention.

As you can see, creative market research does not have to be elaborate, expensive, or time-consuming in order to be effective. By using your creativity, you can easily gather and assess information that can be vital to the success of your business.

HOW DO I EXPAND MY PRODUCT OR SERVICE?

We used McDonald's restaurant as an example of a business that successfully found and developed its market niche. Because of the intense competition in the fast-food business, McDonald's is constantly updating its products. When their research told them that consumers would buy breakfast foods, McDonald's made extensive changes in order to expand their niche. Not only did they test and add new products (breakfast foods), they increased the hours that their restaurants were open. They hired additional staff and modified their cooking methods. The transition was costly, and was undertaken only after their market research told them that it would be profitable.

As a second example, consider the success of salad bars in fast-food restaurants. Today's society is much more health-conscious than it was a few years ago. Adults are concerned about nutrition and restaurants have responded to that concern. Successful businesses like McDonald's constantly monitor their customer's "profile" and adapt their products to the changing demands of the consumer.

Business managers must be sensitive to:
- Opportunities to expand based on goals.
- Changes in consumer demands.
- Maintaining a competitive edge.

Our economic system is structured so that goods and services are provided by individuals and groups who own and control production. Businesses compete with one another for profit. This system permits you to become an entrepreneur, but it requires that you stay on your toes, alert to your competition. You can establish a niche in a market and then make it grow, refining and adjusting it to the ever-changing marketplace. A successful business manager sets goals, and watches carefully for opportunities to expand the business based on these goals. Perhaps your niche, or your "piece of the pie," will grow, as shown in Figure 8-4.

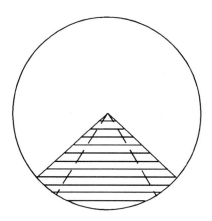

Figure 8-4 Expanding your niche. (Courtesy The Three Entrepreneurs)

A successful business manager is also constantly aware of changing consumer demands and is prepared to adjust the business accordingly. Products or services do not have infinite lifetimes. As new technology develops, older ways of doing things become obsolete. You may establish a market niche, only to find that it changes a few years later and your future success depends on an offshoot product or service that you develop with changing trends. The illustration in Figure 8-5 shows how your niche can change.

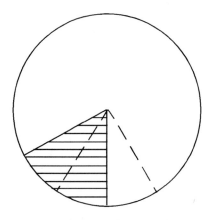

Figure 8-5 Adjusting your niche. (Courtesy The Three Entrepreneurs)

HOW DO I DEVELOP NEW ACCOUNTS?

In order for a business to develop and grow, it needs to constantly generate new business. There is a detailed discussion about attracting new customers in the chapter that deals with advertising. That approach assumes that your customers will come to you once they are aware of the benefits of doing business with your company. There is another complete chapter (24) on developing customers.

What happens if you operate the sort of business in which it is beneficial for you to contact potential customers personally rather than through advertising? This is true of wholesale companies, which service fewer customers than retail businesses but on a much larger scale. In this type of business, it is necessary to keep an account record on file for each customer. By increasing the number and size of the business accounts in this file, an improvement in sales volume can be realized. This is as beneficial as a comparable improvement in material costs, inventory management, or productivity, and in the long run, can prove to be even more valuable.

You have invested a considerable amount of your mind, money, and muscle resources to get your business off the ground. Customer acquisition also involves an investment from you. The development of new accounts can be approached systematically. The method outlined here will not only prove to be efficient in the beginning, but will also be a powerful tool for comparison of profitable accounts in the future. Let's assume you plan to go out and contact potential new accounts. Where do you begin?

Picture the Customer

First of all, you need to form a clear picture of the type of customer you want. Will you be delivering products to a company? If so, you probably want to contact companies that are within a specified distance of your warehouse. Will you need to establish a minimum order size for sales to be profitable? How large does a company need to be before it would be a profitable customer for you?

Locate Potential New Accounts

Secondly, you need to locate potential accounts. One excellent source is telephone directories. These provide a convenient reference of all the businesses in your area that might purchase your product. By noting their locations on a map, you can quickly identify the areas in which potential customers are concentrated.

Screen Potential New Accounts

List your potential new accounts and rank them according to the quantity of your product that they might use. You can then begin the process of calling upon the companies that will potentially become some of your most profitable accounts.

Put yourself in the place of your buyers. If you were shopping for your product, what qualities would attract you? What can your business offer that would convince the purchasers to buy from you instead of from their current suppliers? Take a look at other products that potential new customers sell and the sales techniques used. What is emphasized? Quality? Service? Price?

If you stress the similarities between your values and those of your customer, you are more likely to capture their interest in your product.

Once you have acquired a new account, handle it with the respect that it deserves. It is much easier to maintain existing accounts than it is to find new ones. If you neglect your regular customers, you give them reason to take their business to one of your competitors.

HOW DO I DEVELOP NEW PRODUCTS?

Another way to expand your niche is to develop new products. This requires innovation at all stages, from the inception of the idea through its production and distribution. There are two important criteria for successful development of new products.

The first is that you evaluate the strengths and weaknesses of your company. Do you have idle plant capacity? Do you have access to the resources that a new product requires? Do you have the management staff to oversee the production and distribution of a new product? Will the production of the product maintain cash flow? How long will you need to finance the venture before it becomes profitable? Can the company absorb the losses if the new product fails?

The second, but equally important, criterion for the decision to introduce a new product is that you correctly assess market conditions. Are you looking at a new market for a new product or an expanded market for an alternative product? Are substitutes for your new product readily available? If so, the market for your product may be very price sensitive. Is the product a fad or is it something that will be in demand for a long time? Is it tied to another product? (For example, the number of replacement tires sold correlates with the volume of new car sales three to five years earlier.) Are people spending their money on products such as yours? You may have an award-winning idea for a luxury item, but don't try to introduce it during an economic recession. Wait until market conditions are right.

Once you have made a careful assessment of both your company's capabilities and the condition of the market, you will be more likely to make the correct decision about introducing the product.

When you are ready to expand your business, the next step is to hire a patent attorney. This person will search existing patents to assure that duplicate or better patents do not exist for your new product. These attorneys also prepare patent applications and file them with the United States Patent and Trademark Office.

While the introduction of a new product is a good way to expand your niche, be aware of the dangers of product liability. In recent decades, manufacturers and sometimes sellers have been held responsible by the courts for faulty production or for misrepresentation of quality.

YOU'VE DEVELOPED YOUR NICHE

Now that you have developed your niche, keep track of your performance. You became an entrepreneur by setting goals. You became successful by reaching for those goals. In order to maintain your success, you need to constantly evaluate your progress. What works? Why? Continue to ask yourself questions and evaluate the answers. Remember, specialization is often a key to success, Figure 8-6.

Figure 8–6 Pick-your-own strawberries have a specialized market niche. (Courtesy DAN-D-Acres)

SELF-EVALUATION

1. What does market research mean?

2. Describe two ways to do market research.

3. Joe owns a gasoline station. Describe a plan for expanding his market niche.

4. Suggest three ways Joe can develop new customers.

5. Which four of the following businesses lost the most financially in the past ten to fifteen years?

   a. gasoline   b. furniture
   c. grocery   d. shoes
   e. clothing   f. toy
   g. auto dealers   h. hardware

6. On a separate sheet, write the classification of market niches next to the auto models named below:

Model Name	Class
a. Ford Tempo	Luxury
b. Toyoto Tercell	Sports
c. Chevrolet Nova	Family

    d. Cadillac   Economy
    e. Ford Probe   High Performance
    f. Honda Civic
    g. Lincoln
    h. Pontiac 6000
    i. Jaguar
    j. Chevrolet Chevette

STUDENT ACTIVITIES

1. Using your mall map identify the businesses that have the best chance to expand their market niches.
2. Select two of these businesses and suggest methods for expansion. Compare your answers with those of the rest of the class.

CHAPTER 9

How Do I Promote
My Business?

OBJECTIVE

To recognize the many methods of advertising and their business importance.

COMPETENCIES TO BE DEVELOPED

After completing this chapter you will be able to:
1. Define advertising.
2. Identify the major advertising media.
3. Describe the difference between public relations and publicity.
4. Describe a marketing strategy.

TERMS TO KNOW

Advertising agency	Communicate
Evaluate	Logo
Promotion	Strategy

MATERIALS NEEDED

A newspaper with advertising supplements.

INTRODUCTION

By this time, you may have developed a brilliant idea. You may be ready to offer a product or service that the public will love. *But*...it's no good if no one is aware of it! You have to introduce your company to the public and convince them that they want to do business with you. You do that by advertising.

In this chapter we will discuss the following points.

- What is advertising?
- How do I advertise?
- What's the difference between advertising and public relations?
- Won't my product sell itself?
- What does advertising cost?
- Should I hire help, or can I advertise my product myself?
- What am I advertising? My product or my company?
- What kinds of advertising are available?
- What's best for my business?
- What is good public relations?

Advertising can be as complicated or as simple as you choose to make it. The point is, you need customers and you have to convince your customers that they need you.

CASE STUDIES

The Case of the Yellow Pages

The Connelly Insurance Agency decided that its special market niche would be farm insurance. As a result, they concentrated their advertising budget on a prominent advertisement in the phone company yellow pages. This display featured a farm scene to help readers recognize the insurance specialty.

The Case of Company Promotions

Phil introduced the opening of Exeter Feed and Grain with a full-page advertisement in the local paper, sponsored by the company whose product line he carried. This was followed by more Exeter Feed and Grain ads in the following weeks. Since the business started with very limited capital, the weekly ads were dropped to 3-inch boxes featuring the brand name of the product line and Exeter Feed and Grain. At special *promotion* times larger ads provided by the company were run in addition to the standard 3-inch ad.

What the Cases Tell Us

- Both businesses worked with limited advertising budgets.
- Neither had a cohesive advertising plan.
- Both depended on word of mouth for growth.
- Franchised companies often provide cooperative advertising in the local area.
- Franchised companies provide ready-made ads that the local outlet can personalize and use.
- People must know about your business before they can use it.

PROMOTING YOUR BUSINESS

What is Advertising?

Advertising is a method of getting attention. It says, "Look at me! Look what I can offer you!" It draws customers to your business and, at its best, can create a desire for something that you can fulfill.

Why Advertise?

Businesses advertise for many reasons. Advertisements create customer awareness. Figure 9–1 shows one way to make people aware of your business. By advertising, businesses attempt to stimulate sales, to establish a strong image, or to change an old one. The bottom line is that they advertise because they want to make money. A well-designed advertising plan will increase profits.

Figure 9–1   Magnetic door signs offer a way to identify company vehicles. (Courtesy The Three Entrepreneurs)

How Do I Advertise?

Advertising is often considered an art. Consumers seldom think about why a new product has caught their attention. If they do stop to consider a certain advertisement, they might think that someone luckily stumbled upon an eye-catching *logo* or a pleasing tune and that it just happened to capture public attention. This is not the case. Advertising has become a science. Businesses spend a great deal on marketing research. In fact, some companies devote a large part of their budget to advertising. They have discovered it pays. Consider several ways you can successfully advertise.

Know What You're Selling.   This may seem simple. Of course you know what you're selling. But consider all the aspects of your business. There are several things that you want to present to the public.

Product	Name
Service	Location
Image	Sales
Price	

If you attempt to emphasize all this information in one advertising campaign, you will confuse the public, create more expense for yourself, and defeat your purpose. One thing advertising professionals have learned is, "The simpler, the better!" You want your advertising to be memorable, to leave an impression.

Decide what it is that you are selling. What gives you an edge over your competitors? Price? Quality? Convenience? Service? The key is to think about your business from the consumer's viewpoint. What is it that will persuade the client to come to you? If you offer a product that no one else in town offers, or if your product is of a higher quality than the competition's, then you may want to emphasize that. If your price is the best around, use that. Perhaps your location is convenient. Maybe you have a high-quality service that you want to publicize, or it could be that your business image is what you want to *communicate* to the public. The approach you will take in advertising depends on what you decide you want to sell.

Identify Your Customers.   Marketing cannot be all things to all people. Study your prospective customers. Ask yourself the following questions. Who are they? Where are they? What appeals to them? What are their needs? What is their average income? How old are they? What are their buying habits?

A billboard in a low-income residential area would be less expensive than one rented along a major traffic route. However, if you're selling recreational vehicles, you would be wasting your advertising investment if you selected a lower income area in which to advertise. People in high-rise condominiums aren't interested in lawn mowers. Those that use city buses for transportation won't call you to fix their transmissions. Teenagers aren't interested in backyard grills and those from the middle-aged group don't purchase many rock albums.

You want to pinpoint the group of consumers that will be interested in your business and find out where they are located. Consider your sales zone, the customers you want to reach. Are you mowing lawns in an area of six square blocks? Or do you want to draw customers from several miles around? Perhaps you're starting a mail-order sales business and want to contact potential customers from across the country. How do you reach them?

Next you need to discover what appeals to them. Do they like a high-class image, or are they more likely to be attracted to a country, home-style concept? Are they shopping for convenience, or would they prefer the atmosphere of a specialty shop? Once you know your customers, you will have a better idea of how to advertise.

Establish a Marketing Strategy.   How much will you spend? To determine your *strategy,* you need to know your approximate advertising budget. Then you will be able to decide what media you want to use. Yellow pages represent one media, Figure 9–2. Remember that advertising is considered a business expense and is tax deductible.

Figure 9–2   Yellow pages advertising helps people locate your business or service. (Courtesy Donald F. Connelly)

Timing is important. You may determine how much you want to spend on advertising in the first year, and then spend most of that as you establish recognition for your business. You will need less later when you begin to develop regular customers.

The rate of growth you want to achieve will also affect the amount you budget for advertising. If you want to boost sales, increase advertising. If you want to maintain your level of sales, advertise only as much as you need to continue your current business activity.

You might want to take a look at what your competition is doing. How are they reaching customers? Is it effective? How much are they spending? What media are they using? Here are some of the ways that businesses determine how much they will budget for advertising.

Percentage of Estimated Sales.   Using this method, you designate a percentage of your total sales over a certain period of time for your advertising account. When you are starting your business, you need to refer to your anticipated sales for the first year, and use a certain percentage of that amount allocated over 12 months. You need to look at your sales after a few months and adjust your advertising budget according to the volume of business you are doing. If you are not selling as much as you had expected, increase your advertising.

Percentage of Units Sold.   Using this method, you allocate a specific amount of your profit from each unit sold for future advertising. For example, your product sells for $2.15. You decide to budget 15% of each unit sold for advertising, and therefore invest $0.32 per unit sold. At the end of a month's time, when you have sold 400 units, you have $128.00 in your advertising account for the next month.

Calendar Sales.   If your business is seasonal, you will want to do your heaviest advertising at certain times of the year. If you are painting houses during warm months, you will want to advertise your business during the early spring and summer months. Conversely, if you are operating a snow removal business, you will want to promote your business in the fall and winter months.

How Will You Advertise?   The decision of how to advertise is no longer a simple one. Many businesses hire expert help from advertising agencies. These agencies offer many advantages. They know the local markets and have researched the effectiveness of different types of advertising for various businesses. They are aware of what's available and what it costs. Advertising packages, using many types of media, can be created for you. They will design your logo, write your copy, illustrate your ads, and place them for you. In addition, they will also help you *evaluate* the results of your advertising. A business pays for the expertise of an accountant or a lawyer. An *advertising agency,* too, is a kind of specialist. It can help a business put its advertising dollars to their best use. The agency can also save you time, which may be in short supply as you work to get your business off the ground.

You may prefer to tap your own creativity. Creating your own advertising program will require some footwork on your part, as well as a lot of imagination. But there is help available. Newspapers employ people who work with you to put together advertising copy. Radio stations also have personnel who help write radio copy. Often manufacturers will help with advertising expenses because local rates are less expensive than national rates. They may be willing to help you advertise in exchange for better shelf position for their product, for example, or furnish in-store display materials.

Design a Media Plan.   There is a wide variety of media options available to you. You may want to test one or two, or you may try several at once. Here are a few ideas.

Newspaper.   You can buy space on a newspaper page or have inserts printed and included with newspaper delivery. Rates increase with the circulation of the paper. Your options range from an advertisement in the classified section to a full-page color layout in the Sunday supplement.

Radio.   You can write a poem, sing a song, or create a dialogue. You can have your advertising copy read with music in the background. Rates depend on the length of the advertisement and on the time of day that it airs.

Magazines.   These usually cover a larger sales zone than newspapers, but you may find a local magazine which depends on advertising for income.

Television.   Your options range from local stations to national coverage, although advertising on the larger networks often runs into hundreds of thousands of dollars. You would probably want professional help in putting together a TV commercial since there are many technical factors that must be considered.

Logos/Slogans.   A well-designed logo can quickly become a part of your business image and can develop immediate viewer audience recognition. It can be displayed on company vehicles, letterheads, signs, and so on.

Direct Mail.   Direct mail lists can be obtained from many sources. You might develop a preferred customer list and mail letters, postcards, announcements, coupons, or invitations to the names on your list. Figure 9–3 shows a direct mail piece.

Figure 9–3   Direct-mail brochures are one way to tell people about your product. (Courtesy The Three Entrepreneurs)

Fliers/Circulars.   Federal law prohibits you from putting anything in other people's mailboxes. However, you can distribute fliers on car windshields, in doors, on store counters, or have your own personnel distribute them at fairs or public activities. A word of warning—fliers often get dropped in parking lots, and that may be detrimental to your business image.

Catalogs.   According to the old saying, "A picture is worth a thousand words." You might want to develop a bulletin, a brochure, or a catalog to mail to your customers.

Demonstrations/Workshops/Trade Shows.   You can rent space in a mall, at a sidewalk sale, or at a trade show and introduce your company to a new set of customers.

Signs/Billboards.   These have become a part of the scenery today, but an attractive and imaginative sign will still capture attention. It's a necessity at your business location.

Window Displays.   Some people are spur-of-the-moment shoppers. A large enticing window display may attract a new customer to your parking lot.

The Yellow Pages.   Don't miss this one. Many people turn to their telephone directories when trying to locate a service or product that they don't use frequently.

Business Cards.   This is a simple way of emphasizing your professional status. Some people collect them. A lot of people use them for future reference.

Odds and Ends.   Many businesses distribute "freebies" or samples. You can have your name and business logo printed on pencils, matchbooks, shopping bags, ping-pong paddles, balloons for children, and so on. Customers remember generosity. Your investment may be returned many times over.

Measure the Results.   This may be the trickiest part of advertising, but it is the most important. You need to know how effectively your advertising dollars are being spent. What is a good investment? What received a small response?

If you're in the mail-order business, a promotional code is a simple way to measure the value of your advertising dollar. Assign a code to each advertising media and have that code printed on your order blanks. When the orders come in, simply keep track of the sources. Before long, you will know which brought in the most customers and the most sales. You can then concentrate your advertising on those sources.

If customers are coming into your business to shop, you can ask them at the sales counter how they heard about your company.

Most importantly, keep good records of your advertising. What media did you use? When did you advertise? What did your sales volume do during the period following a certain advertisement? What works and what doesn't work? Analyze your marketing investment and use the results to create better and more effective advertising for your business.

Publicity

Publicity is information about your business and your products that captures the public interest. Publicity makes people notice you—it puts your company in the public eye and can be good or bad. For the sake of your business image, which we will discuss a bit later in this chapter, you're interested in the good publicity. There are many ways you can stage publicity for your business.

Open Houses.   Have a grand opening or an anniversary celebration. Invite the public into your establishment and then make them feel welcome and comfortable. This would be a good time to pass out "freebies." You might offer them a refreshing drink, a treat, or a tour of your new facilities. Once they become acquainted with you and your business, they're likely to return. Some of them will become regular customers.

Community Events.   Your business publicity can be present wherever and whenever the public gathers. Have displays set up at community fairs. Sponsor a Fourth of July fireworks display or a Christmas tree decorating contest.

Civic Activities.   Develop a connection between your business and community pride. Sponsor a play, a festival, a bicycle race, or a baseball team. Develop a safety awareness program.

Charitable Organizations.   Contribute to local clubs. Support medical research. Sponsor projects put together by a group of retirees or a home for mentally handicapped people. Then make sure the public is aware of your efforts to better society. Display posters telling your customers that you contribute a nickel to charity for every dollar they spend at your business. Call the newspaper and tell them of your interest in the community and the contributions that you have made.

Word of Mouth.   Your best publicity will come from satisfied customers. If someone expresses satisfaction with your business, don't hesitate to tell them that you'd appreciate a recommendation to their friends and neighbors.

Public Relations

Your reputation will be one of your most valuable business assets. Although it can't be measured financially, it will affect your customer base, your employee satisfaction, and your business relationships. Think about the type of image you want your business to present to the public.

Business Image.   What impression do you want people to have as they think about your business? Do you want people to think of your company as fair? Responsible? Clean? Exclusive? Your management will have a major impact on the creation of your business image—an image not only for your customers, but for everyone with whom you deal.

Your Community.   This includes your neighboring businesses. Are they glad to have you on the block? Will you cooperate for sidewalk sales? Shoveling snow in the winter? Dealing with emergencies such as water problems when a water main breaks?

Your Company.   This includes your employees and their families. Are they happy to be working for you? Are you supportive? Your employment policies have a major impact on their lives.

Your Trade Associates.   These include your suppliers and distributors. Will you deal professionally with them? Fulfill your commitments?

Your Financial Connections.   These are your banker, your investors, your stockholders, your insurance agent. Will they see you as trustworthy? Responsible? Dependable? Will you be one of their preferred clients?

Good public relations begin at your front door. As manager, it will be up to you to establish your business policies. Here are some of the things you will need to think about.

- How will your sales people conduct themselves?
- What sort of customer service policies will you develop?
- How will you handle customer complaints?
- How will you handle unusual requests?

You will want to develop your own public relations philosophy. Remember that your policies affect not only your customers, but everyone else that you deal with, from your employees to your banker.

SELF-EVALUATION

Promoting Business Quiz

True-False: On a separate sheet, write T for true statements and F for false statements.

1. It is important that your financial connections have a trusting relationship with your business.
2. The public image of a business is dependent only on the owner.
3. Your reputation is important to your business.
4. The reputation of your employees reflects on your business.
5. Businesses should contribute to all local causes.
6. Word-of-mouth advertising is of little importance.
7. Special promotions, such as an open house to launch a business, are important.
8. You don't need to formally evaluate your promotions as sales figures tell the story.
9. Yellow pages are an important media to generic businesses.
10. Logos are of little advertising value.
11. The service of an advertising agency is worth considering even within limited budgets.
12. Most entrepreneurs will have enough creativity to handle their own advertising campaigns.
13. An advertising budget based on units sold makes sense after the initial start of the business.
14. The advertising budget should reflect your desired rate of growth.
15. An advertising campaign is properly geared to the general public.
16. Billboard location matters little if you decide to use that type of advertising.
17. The timing of an advertising campaign is very important.
18. An advertisement needs to give all the details.
19. Customer awareness is a common goal of advertising.
20. Franchised companies frequently provide help to local dealers.

STUDENT ACTIVITIES

1. Pick a business in the local area and clip all advertisements from the newspapers over a period of time. Summarize the ads and evaluate advertising versus patronage. Is there an observable correlation?
2. Do the same for radio, T.V., or other media.
3. Plan an advertising campaign for your business plan.

CHAPTER 10

How Do I Account For Business Transactions and Events?

OBJECTIVE

To identify the concepts of business transactions and events.

COMPETENCIES TO BE DEVELOPED

After completing this chapter you will be able to:

1. Describe what an accountant does.
2. Explain how the approaches of entrepreneurs, appraisers, economists, and accountants differ.
3. Define basic accounting terms.
4. Explain basic records needed by a business.

TERMS TO KNOW

Accounting rules	Backup
Business entity	Cash basis
Categories	Consume
Continuous life	Convert
Cost	Domain
Entity	Estimate
Expected	Forecasting
Invest	Posting
Personal value	Recipe
Present market value	Utility
Records	Wealth
Ward	

INTRODUCTION

Keeping *records* takes time and effort, but it is time well spent. Many people have negative feelings about this, and think the time would be more profitably spent in conducting their business. How do we discipline ourselves to keep records until our business prospers and we can hire someone to do it for us? We need to develop a positive attitude toward what these records will do for us and for our business.

Read the following case study about Jerry and his friend and their lawn buildings. Where would they have been without their records when a new estimate was needed? It would have increased the time needed to complete the estimate. Records save time and money. They must be kept up-to-date and orderly. Doing this can give us a feeling of worthwhile accomplishment.

We need records to:

- know our cash balance.
- know who is in debt to our business.
- keep track of our business debts.
- file our tax reports correctly and promptly.
- know our financial position.
- help make business decisions.

CASE STUDIES

The Case of the Helpful Records

Jerry and a friend built several lawn storage barns and realized a good return on their work. The following spring another neighbor asked the boys to build one but wondered if it could be done at a reasonable price. He wanted a slightly larger building than the ones they had built the year before.

Since Jerry was by nature a methodical person, he was able to look at the cost of the original lumber, calculate the extra material needed, and then obtain current prices. He knew how many hours it took to build each of the original buildings so he could estimate the extra time they would need for the larger model. His estimate was prompt and the sale was completed.

The Case of the Missing Records

Late in December, Phil was talking with a customer at the front of the store when Steve, the new assistant, came up to Phil and said, "I can't find Ned's account in the accounts receivable ledger. Can you help me?"

Ned had come in to pay his bill before the first of the year as he reported on the *cash basis*. Leaving Steve to help the customer, Phil excused himself and went to handle Ned's account. Phil had some of the accounts out of the ledger at home where he was *posting* recent purchases.

The Exeter Feed and Grain Company used the multiple ticket register in which one copy of the sale is given to the customer, one to the bookkeeping file, and the third copy is filed. Knowing the account was current except for two grain sales and one feed sale, Ned and Phil were able to pinpoint the dates of the deliveries, locate the duplicate copies, and settle the account.

With the amounts owed ranging from $400 to $600 each, it is easy to see the importance of an adequate records system. The missing account sheet would have been serious without a *backup* record system like the file copy of original sales tickets.

What the Cases Tell Us

- Jerry's records of costs were time savers when a new estimate was needed.
- Records serve as planning resources as well as tax information.
- Detailed records are an asset.
- Loss of records could be costly.
- A good records system has a backup.
- All clerical employees need to know how accounts receivable are handled.
- With a computerized system the records would have been instantly available.
- With a computerized system the multiple posting would have been completed with the original entry.
- Records are an aid in decision making.

ACCOUNTING FOR YOUR ACTIONS

The Business of Living

Learning, earning, consuming, and *investing* are words you might use if you were to describe, somewhat in business terms, the processes of living out your life. You come into this world with no material resources and only a potential ability for developing your mind and muscle resources. Thanks to the influence of your parents and others, you have survived and you have learned significant mind and muscle skills. Perhaps you have even received gifts of material resources. Certainly, other persons consumed some of their resources to provide for your living to date.

From now on, you must know something and you must do something, unless you expect to be a *ward* (or in the care) of others. Learn all you can; you can use your mind and muscle skills again and again. Earn what you want; you need to earn enough to survive and you want to learn enough to have a meaningful life.

As a creative young entrepreneur you will recognize opportunities to use your resources to earn money. You do this when you are able to create new products or services which you sell at a profit. By this process, you *convert* some of your mind, muscle, and material resources into earnings.

You earn this new money resource and you use it in your business of living. You buy and *consume* material resources that permit you to survive and to have a meaningful life. If you earn more money than you choose to consume immediately, you save the extra money. You add it and the resources you buy with it to the material resources or *wealth* you previously had.

For much of your life, you will likely earn more money than you will consume. The amount of your material resources, and thus your financial wealth, will increase. You will use these new resources in your own business, *invest* the money in others' businesses, or save it and loan it to others in exchange for interest payments.

Later, in your business of living, you may consume more money than you create. Your wealth will decrease. You may decide this is OK as long as your money and other material resources stay above zero.

Other people in this world will be better off because we lived here. Most of us will still have some material resources to give to the next generation when we die. We will have already helped our children survive and shared our mental skills with them. When we die, they will likely have more mind, muscle, and material resources than we had at their age.

Accounting for Your Mind, Muscle, and Material Resources

How can you account for or measure the amount of your present resources? Is that important? Correctly measuring your current mind, muscle and material resources is the first step in the management process of using what you now have to get what you want. Without knowing what resources you have, you might try to do things for which you are not well-equipped, or you might waste resources that you didn't realize you had.

On any date, you can take a series of mental exams and physical strength, dexterity, and speed tests, and you can make an inventory of your material resources. By these processes you can measure what resources you now have.

On a later date, you could take similar tests and make another inventory of material resources. The differences in your resources would represent the changes which occurred during the period between the dates because of what you learned, earned, consumed, and invested. By this process, you could account for your performance for this period of time.

As you thought about what you did, you would be able to identify things you did well and things you would want to change if you were ever to do the same types of things again. For most people, past performance is the best estimate of future performance. Consider it carefully as you decide what to do. Remember, of course, that you have the potential to learn from your past experiences and to change your performance in the future.

You begin life with the potential for learning and earning. You learn both muscle and mind skills. You learn all these skills from others. Up to now, some of us have had better teachers and/or been better students than others. From now on, each of us can use what we now have to get what we now want most. We can learn and thus we can improve our mind/muscle skills and our performances.

For example, you can learn and use new skills such as entrepreneurship, marketing management, production management, personnel management, financial management, accounting, etc. That's exactly what you'll do in this activity.

Deciding What You Want To Do

Management is using what you now have to get what you now want most. Rob's management of this lot of steers will help him reach his goals, Figure 10–1. Some persons do a better job than others. With about the same resources, they realize more satisfaction in life. They are better managers.

You can maximize your satisfaction if you do the right things with your inventory.

Figure 10–1   Accurate records are important to Rob's success with his steers.
(Courtesy Kathy Shanks)

Present Personal Value of Resources

You must determine your present *personal value* in using each of your resources. You do this by thinking of all the possible ways of using your resources. Then you pick the ways that you think will be best.

You are now learning about resource present personal value. Consider this thought. You keep a material resource whenever it's worth more to you than the satisfaction you would realize if you were to sell it and use the money in another way.

Present Market Value of Resources

In our economy, people exchange their resources with others for money or they trade for other material resources. Prices are agreed upon as people bid against each other in the marketplace. Prices are affected by the demand for and supply of resources, including money. Persons called economists study how prices are affected by people as they conduct their daily business. You will learn about how you can use some of their conclusions in another chapter.

As you consider the use of your resources, you estimate the market prices you expect to pay for wanted resources and the prices you expect to get for resources you might sell. In this process, you estimate the *present market value* of resources.

Persons called appraisers are experts at estimating the likely price different types of resources would command if sold. Often, other persons hire appraisers to *estimate* the present market value of a particular resource before they decide to buy or to sell it. As an entrepreneur, you use appraisal skills when you decide what to make and/or market.

Accounting for Changes in Amount of Your Resources

When does the amount of your material resources change? Assume you can measure your material resources by putting them in different *categories* and then can count each category. On any date, you can find the difference that has occurred since the last date you made a count. By this process, you've accounted for the change in the physical amount of your material resources.

A good set of records can help you with accounting, Figure 10–2.

Figure 10–2 Dave and Tammy Buck recognize the value of records. (Courtesy David Buck)

Accounting for Changes in the Value of Your Resources

Now, let's consider a harder question. How can you account for the changes in the value of your material resources? The question would be easy to answer if you started and ended with only money material resources. The difference in money between the beginning and ending dates is the amount by which the value of your material resources has changed. Assuming the value of money is still the same, you could tell the difference in market purchasing power you have now compared to the market value of the money you had earlier.

The question is much harder to answer when you have other material resources. Because the demand and supply of these material resources actually changes, the market prices change. Therefore, until a resource is actually sold, it's hard to estimate what the selling price will be. This problem has caused persons to develop different solutions to the question depending on their needs and uses for the information.

For example, you need an estimate of present personal value and present market value as you decide how to use your resources. However, neither of these estimates is very objective. Different people, looking at the same information, would likely have different estimates. Professional accounting practices and rules have been developed by persons who want to help others communicate information about financial transactions between entities and events within entities. Partly because of the problem of objectively making estimates about future transactions and events, professional accountants generally record only transactions and events that have already occurred.

In the next few paragraphs, we will present the important features and uses of the present personal value and present market value concepts. Then we will present professional accounting practices in more detail. You will use this information in choosing how best to use your resources to achieve your goals.

Who are Accountants?

Business records are the *domain* of people known as accountants. Their task is to maintain needed business records and reports. This record keeping is called accounting.

Accountants assume that a business is a separate *entity* from the owner of the business. (Webster defines an entity as something that exists separately or independently.) In a general way this is true for business. We will consider this to be the case for this discussion.

Who are accountants? They are the financial historians of a business. They record transactions and events associated with the *business entity.* Personal transactions and events will be considered later. For the moment we need to understand simple accounting since most of us will need to provide for our own record keeping as we start to carry out our business plan.

Accounting for Resource Personal Value

Suppose you were to estimate the present personal value of each of your resources. Then on a later date, you made the same estimate. Any difference between the estimates on the two dates would represent the change in the present personal value of your resources during that period.

Accounting for Changes in Resource Market Value

Selling each of your material resources is one of the recipes you have in your decision-making process. You will want a market value appraisal of each of your resources because on any date you might decide to sell one or more of them.

When you exchange money for another material resource, you buy it. However, you may not be able to sell it for the same amount of money. Therefore, the market value may be different from the price you paid. (In a later section, you will realize that accountants have chosen to exclude this fact from their formal *accounting rules.*) It is important to realize that on a different date, the demand for and supply of resources, including money, may be different.

As you will discover, lenders also want to know the market value of your resources. If a borrower quits repaying his loan, lenders want to be able to get their money back by selling the borrower's resources. Therefore, lenders want a current estimate of the appraised market value of your resources to secure your loan.

The Selling Decision

When do you decide to sell a material resource? That is, when do you decide to exchange it for money? You decide to sell when the present personal value of the resource is less than its present market value. After considering all the possible recipes for continuing to use it, you conclude the best *recipe* is to sell it.

Note that the selling decision is not affected by the *cost* of the resource. Cost is past history. You purchased the resource at a time when the market demand and supply, and thus the price, were different. You make the selling decision based on your estimate of present and future prices.

The Entrepreneurial Decision

How do you decide to use some of your resources to make and/or market other resources? You can decide to make or buy a new resource when you think you can sell it at a profit. But that's just part of the answer. You may find recipes for making more than one resource at a lower cost than its *expected* selling price. How do you select the best recipe? Look at all the combinations of recipes and select the one that you think will be best.

Consider *expected* resource costs and *expected* selling prices, and then decide how to use your resources to get what you now want most. This decision process is very old. People used it long ago, even before our ancestors created our present market economy or our present accounting practices.

People have developed explanations to describe the effects of choosing how to use resources to achieve goals. These explanations are called ''economics.'' Economics is a description of decisions that most persons have made in the past. By understanding these concepts, you can predict how people will make decisions in the future. By learning about how people generally act as they decide to use their resources, you will be able to make better decisions about how to use your resources. You will learn more about economics later.

Our present accounting practices were also developed over time. These practices are a system for describing actual transactions between persons or other entities. Today almost everyone uses accounting terms to describe their actions. They read accounting reports to find out what happened, financially speaking, in a business over a recent time period. You will learn about accounting next. Many businesses use computer accounting software to keep records, Figure 10–3.

**Figure 10–3   Bob demonstrates his computer records system for a student.
(Courtesy Robert L. Maudlin)**

Professional Accounting Practices

The question, ''How can you account for changes in the value of your material resources?'', has another set of answers. These answers have evolved in response to many persons' needs for objective financial information about businesses. Accountants have developed rules and generally accepted practices for keeping historical records of financial transactions and events.

(Since accountants use only historical records, the word "value" in the question should be replaced by the word "cost." Value as used in present market value and present personal value is an estimate of expected future transactions and events. These activities might not occur. Since accountants don't recognize transactions or events that haven't happened, you will want to hire an appraiser or develop your own economic *forecasting* skills.)

Accountants work mainly for businesses that are owned by other persons. They are concerned about accounting to determine:

1. Where the money resources came from to start the business.

2. How much earnings were created by the business for each reporting period.

3. How much money resources were returned to the owners of the business each period.

Answering questions 1 and 3 is easy. Question 2 is more difficult. We will help you learn how accountants answer that question in the next section. Then, you can use accounting rules to make reports for yourself and others. Also, once you know the rules, you can understand reports that accountants have made for other businesses.

As a general practice, accountants choose to recognize changes in values for material resources only when a sales transaction occurs. Therefore, between sales, a resource is recorded by accountants at its cost to the resource owner, and not at market value or personal value. Further, if a resource is partly or completely consumed while making a new resource, the cost of the original resource becomes the cost of the new resource.

The professional accountant's practice of waiting until after a sale has occurred to recognize a change in resource value is quite safe. An accountant reports only historical facts. This means that resource market value or present personal value estimates are not reported. Even though changes in demand, supply, and market price are known, an accountant chooses to record only transactions between entities and events within entities that have occurred. (As an entrepreneur, you develop estimates of value based on your expectations of future transactions and events. You may even hire economists and/or appraisers to help you.)

Accountants assume business entities have *continuous life.* This assumption is consistent with their practice of assuming that the purchase price of a resource is its present cost to the entity. Since accountants assume continuous business life, they are not concerned about the market value of resources that might have to be sold prematurely. (As an entrepreneur, you are concerned about this possibility. We will help you account for it later.)

You will use information based on professional accountants' rules as you decide how to use your resources. Therefore, we will present many of the generally accepted accounting practices in the next section. Remember that you also need resource present personal value and present market value estimates for your decision making. After studying some formal accounting, you will learn to use the terms, where appropriate, as we help you develop records and reports for use in decision making.

As you study accounting, it may help you understand the material if you consider Figure 10–4, Similarities and Differences in Approach to Specific Issues by Entrepreneurs, Economists, Appraisers, and Accountants. This will help you understand how financial matters are approached from different viewpoints.

SIMILARITIES AND DIFFERENCES IN APPROACH TO SPECIFIC ISSUES

	ENTREPRENEUR	ECONOMIST	APPRAISER	ACCOUNTANT
Distinguishing characteristics	Makes decisions for using resources and carries them out in both business and personal life. Accepts most of the risk and gets most of the reward, both gain and pain, from the efforts made.	Observes buying and selling performances of other persons. Develops descriptions of how people generally act in the market under various circumstances. Based on observations of past performances by people in the market, makes forecasts about likely future performances by people in the market.	Uses economic information to develop an estimate of how a typical person would use a specific resource or group of resources.	A financial historian. Develops a set of accounts and reports for recording, summarizing, and describing what happened, financially speaking, within an entity and between this entity and other entities.
How material resources are measured	Present Personal Value. This is present market value for each resource plus the personal *utility* or satisfaction gotten from having resources in their present time, form, place, and possession.	Present Market Value.	Present Market Value. For each resource or set of resources, the appraiser estimates value three ways as evidence of the final conclusion: sales prices of similar resources; earnings if the resources were to be used by a typical entrepreneur; replacement cost.	Past cost to entity. The total cost of the material resources is shown as the dollar amount of the assets on an entity's balance sheet.

SIMILARITIES AND DIFFERENCES IN APPROACH TO SPECIFIC ISSUES

	ENTREPRENEUR	ECONOMIST	APPRAISER	ACCOUNTANT
How claims are measured	Past cost to entity.	Past cost to entity.	Past cost to entity.	Past cost to entity. Lender claims due listed as liabilities. Owner claims are listed as net worth on an entity's balance sheet.
How past performance is measured using balance sheets	Consider net worth change plus other utility or consumption realized.			Calculated entity's net worth change plus dividends distributed in the period between two balance sheets.
How past performance is measured using an income statement	Income statement plus other non-money utility or satisfaction.	Income statement of typical entity.	Income statement of typical entity.	Income statement of this entity.
How future performance is estimated	Projection of profitability and other utility performance including consumption versus wealth creates decision.	Project profitability performance of typical entity.	Project profitability performance of typical entity.	Projection of profitability performance for use within an entity only, not for distribution to stockholders or other outsiders.
When earnings are recognized	When entrepreneur decides to follow a specific set of recipes for using resources based on expected costs and sales revenues.	Interpretation by economist of expected costs and sales revenues of typical entity.	Interpretation by appraiser of expected costs and sales revenues of a typical entity using a specific set of resources.	When earnings actually realized, *i.e.,* (when a sale is made).

CHAPTER 11

How Do I Use a Single Entry System?

OBJECTIVE

To identify the elements of a single entry bookkeeping system.

COMPETENCIES TO BE DEVELOPED

After completing this chapter you will be able to:
1. Recognize the basic entries that define a single entry system.
2. Identify a proper basis for handling small or petty cash purchases.
3. Identify the advantages of computer record keeping systems.
4. Identify the importance of a daily business summary.
5. Identify the types of records that should be preserved from fire or other hazards.
6. Describe the advantages of a separate business bank account.
7. Correctly identify the period of time records should be maintained and preserved to meet legal standards.

TERMS TO KNOW

Accrual basis

Double entry

Paper trail

Preservation

Revenues

Single entry

Cash basis

Microfilm

Petty cash

Protected

Sales slip

Statutes of limitation

INTRODUCTION

The simplest bookkeeping system has been called the *single entry* system. This is because the entry is made only in the checkbook or the checkbook and the journal. The *double entry* system, which contains detailed debit and credit sides of each entry of a transaction, will be covered in the next chapter.

CASE STUDY

Lilly opened her "party makers" supply business in a small city. She was faced with the need to set up a record keeping system. Since Lilly and her sister were the only employees, a simple system would provide the information they would need to file their income taxes accurately. They decided that they would use a *sales slip*, Figure 11-1, for every sale and run all money through the checking account.

What the Case Tells Us

- Simple records can meet all legal requirements.
- A record of all revenues and expenditures is needed in order to properly document tax returns.
- Some system of handling purchases too small for a check is needed.

MAKING RECORDS SIMPLE

The use of single entry records is the simplest record keeping system that is adequate to provide tax records. A checkbook and a sales slip book can document *revenues* and expenditures. With a little more effort, a journal with each sales slip and each check recorded will prepare a better record.

This single entry system must provide the business owners with:

- the assurance that they will pay only the taxes they owe.
- prevention of errors in accounts receivable.
- prevention of errors in payment for goods and services.
- an up-to-date cash position.
- a possible projection for the future.

We want to stress that the double entry system featured in the next chapter does provide an extra measure of checks and balances that keeps the books accurate and balanced with a minimum of extra effort.

Deciding on the System to Use When Setting Up Your Books

Among the considerations in this task are the following questions:

- What reports to ownership are required?

- What experience do the operators have in bookkeeping or are funds available to have the bookkeeping done for you?
- How much money is involved over the course of a year?
- Will the business have access in the near future to the use of a computer?

R. J. Marshall
Retail Sporting Goods Inc.
18 E. Martin Drive
Highland, NY 12528

_____ 19 ___

Name _____

Address _____

QTY.	DESCRIPTION	UNIT PRICE	AMOUNT

Figure 11–1 Lilly decided that a sales slip for each transaction would provide an adequate record system.

Reporting to Owners.   When you only have to report to yourself and to the IRS, you have the choice of using the simplest of records systems, Figure 11–2, with sales slips for revenue and the canceled checks for expenditures or a journal and checkbook system. On the other hand, if reporting to multiple owners is involved, the more detailed double entry system is advisable because of the built-in debit-credit entries that maintain books in balance and at the same time provide a superior *paper trail* for the accountants who examine corporate accounts.

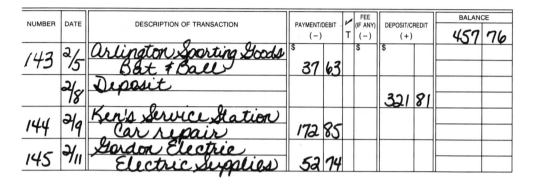

NUMBER	DATE	DESCRIPTION OF TRANSACTION	PAYMENT/DEBIT (−)	T	FEE (IF ANY) (−)	DEPOSIT/CREDIT (+)	BALANCE
							457 76
143	2/5	Arlington Sporting Goods Bat + Ball	37 63				
	2/8	Deposit				321 81	
144	2/9	Ken's Service Station Car repair	172 85				
145	2/11	Gordon Electric Electric Supplies	52 74				

Figure 11–2 Your checkbook register represents a single entry system if all money is deposited and checks are written for every expenditure.

Experience in Bookkeeping.   The sole proprietor or partners starting a new business enterprise might decide on their bookkeeping system based on their knowledge and experience with accounting procedures. Experienced operators will most likely set up the more extensive system to take advantage of the processed information this will provide. If they lack experience and knowledge, they may opt for the simple form in the beginning and grow into the double entry system as they learn more or earn enough to hire someone to do the accounting.

Many small businesses have had an experienced accountant set up a set of books for their business. They then have the accountant teach them how to do the records with occasional help when needed. Many other small entrepreneurs have used a sales slip system and have had a private accountant post the items on a week-by-week basis. In such cases the ticket system should provide multiple copies—one for the customer, one for the accountant, and one to keep at the business.

The Volume of Business Handled.   If the volume of business handled is small, the single entry system is more appropriate than it is in a large volume business. Each transaction generates a document to be recorded. The larger the volume of documents to be accounted for in the financial reports, the more important a better system will become. The advantages of double entry systems suggest its use in high volume situations.

Will Computer Accounting Be Possible?   If a computer with a suitable software accounting system is available, the more sophisticated double entry system is the logical choice. The computer can record entries in the right places and update accounts with a few key strokes. The computer program can also generate many reports, doing in minutes what would take a clerk several hours to accomplish.

Should You Use the Cash or Accrual Basis for Accounting?

The single entry system is essentially the cash method of accounting because the payments for goods and services are recorded when they are made rather than when the obligation is incurred. In the accrual method, the expense is recorded when incurred regardless of when it is paid. In *cash basis* accounting the revenues are recognized when payment is received; using the *accrual basis,* the revenues are recognized when the credit is extended or the charge is made.

Either method is accepted by the IRS, but once adopted, the system cannot be changed without special permission from the IRS.

Installing the Records System

There are some basic details that an entrepreneur needs to consider in establishing a records system. Among the first is obtaining the basic forms. These are available from many printers and systems are marketed as a package by office supply stores. When the particular system has been determined, everyone who writes sales tickets needs to use the proper forms and complete them accurately. No records system can function properly without accurate information properly entered.

Petty Cash.   Most businesses use small sums of money called *petty cash* to purchase small items. Items costing a dollar or less are not suitable for check writing. Instead, the cash receipts are placed in the petty cash drawer and these are filed as reference for the check issued.

Daily Business Summaries.   At the end of each day's business, the sales slips should be summarized, the cash counted and the books balanced. It is easier to discover errors when this is done on a daily basis. Any cash overage or shortage should be recorded and isolated. Sometimes a small overage shows up when someone fails to pick up a few cents change or a shortage might be resolved when the missing coin is found on the floor.

Monthly Summaries.   A monthly worksheet is an important intermediate step between period reports. When a few minutes are taken to post each day's summary on the monthly worksheet, the end of the month report takes minutes instead of hours.

RECORD PRESERVATION

Two concepts are important to recognize in this section. Records should be *protected* from fire or theft and preserved over the time of the appropriate statute of limitations.

Preservation of records in case of a natural disaster is the first of the two concepts named. Few businesses ever recover fully from a fire that destroys their records. This is especially true when there is a large amount of money tied up in accounts receivable. Cash in the bank is available after the fire and accounts payable can be rebilled. They present less problems than accounts receivable, which cannot be accurately reconstructed.

Record preservation over a period of years is mandated by tax records, *statutes of limitations,* and historical value. For a business these periods of time may vary from five years to forever.

Income tax records are seldom needed beyond six years although all statutes of limitation expire after 11 years. Good sense dictates that the actual returns should be retained and only the sales slips and other documentation disposed of. Checks and bank statements are normally kept up to seven years as are payroll records and other business detail records. Journals and ledgers should be maintained as permanent records. Some businesses have their records stored on *microfilm* to reduce their storage space requirements.

Setting Up the Bank Account

One of the first activities of the new business should be establishing a business bank account. The reason seems obvious but needs to be stated—business accounts should be separated from all personal accounts.

Not only do tax records demand such action—every accounting system is based on the idea of a separate business bank account. Mingling personal accounts with business accounts will lead to many unwanted problems. A business checking account will provide a way to prevent these problems. When owners draw money from the business for family living expenses, these funds should be drawn by check to provide the accounting documentation.

SELF-EVALUATION

Single Entry Accounts Quiz

True-False: On a separate sheet, write T for true statements and F for false statements.

1. Double entry records are simpler than single entry records.
2. Either single or double entry systems are acceptable to the Internal Revenue Service for income tax reporting.
3. A petty cash fund is a reasonable solution to small cost purchases.
4. The authors suggest that the double entry bookkeeping system is preferable for a business that reports to multiple owners.
5. The single entry system lends itself to reports for decision making.
6. Once a records system is adopted, changing to a different one is simple.
7. The double entry system is known as the cash reporting system.
8. A daily business summary is a useful bookkeeping tool.
9. Better cash control accounting is possible with the daily business summary.
10. All records should be preserved for a minimum of 10 years.
11. Protecting records from fire is especially important for accounts receivable.
12. Cash basis income tax reporting recognizes expenses when the payment is actually made.
13. A separate checking account is only necessary for a business with multiple owners.
14. All records can be destroyed after the statutes of limitation expire.
15. The statutes of limitation vary from 5 to 11 years.
16. Income tax returns probably should be kept permanently.
17. The amount of information on a sales ticket may vary by type of sale such as cash, check, or charge.
18. Microfilm is used to reduce the volume of records kept.

19. Computers can substitute for all paper records.
20. Records are of value only to meet income tax requirements.

STUDENT ACTIVITIES

1. Use Kermit's sweet corn project (page 140) to make entries by using the Journal and Checking Account.
2. Use a local enterprise to make typical single entry records.
3. Create an appropriate set of SOEP records for your use.

CHAPTER 12

How Do I Use the Double Entry System?

OBJECTIVE

To practice the double entry accounting system.

COMPETENCIES TO BE DEVELOPED

After completing this chapter you will be able to:

1. Demonstrate an understanding of accounting terms by selecting the best definitions in the teaching situation.
2. Correctly enter sample double entry items on the provided accounting forms.

TERMS TO KNOW

Business entity	Credit
Debit	Journal
Market value	Net worth
Nonbusiness entity	Personal entity
Precise	

INTRODUCTION

Accountants assume a business is an entity separate from the owners of the business. Since this is true for many businesses, we will also make this assumption for now. Accountants are financial historians for a business. They record transactions and events associated with a *business entity.*

CASE STUDIES

The Case of Missing Knowledge

When Phil helped start Exeter Feed and Grain, the budget provided funds for one employee in addition to Phil. George, the first employee, knew nothing of bookkeeping. As a result, Phil had to set up a bookkeeping system.

Phil's training had included no work in accounting. He started by having Sam show him how entries were made in the books of the parent company from which Exeter grew. Shortly thereafter Phil attended an XXXX Company training session on bookkeeping and decided their system would fit Exeter.

Phil began his accounting procedures by purchasing an invoice machine that provided multiple forms. One copy of each invoice or sales ticket was presented to the customer, the second copy was placed into the accounting file (for bookkeeping entries), and the third filed in chronological order as a permanent record.

Next, Phil needed to train each person active in the business that a ticket was to be written for each sale. Therefore, records would always be available to show what had transpired.

What the Case Tells Us

- Someone in a business should know basic accounting.
- For records to stay complete and accurate, everyone must follow company rules.
- Two characteristics of a desirable accounting system are completeness and accuracy.
- Packaged accounting systems are often available from suppliers at reasonable costs.
- Phil adopted an existing accounting system recommended by the XXXX Company.

ENTITIES

The Three Types of Entities

Accountants recognize these three types of entities: business entities, *nonbusiness entities,* and *personal entities.* Business entities are formed to make money and are known as for-profit businesses. Once formed, accountants assume businesses have continuous life. They may organize as corporations, which have one or more stockholder owners. Dan-D Acres in Figure 12–1 is an example of this. Businesses may be organized as partnerships with two or more other entities as owners. A sole proprietorship with one entity as owner is a third way of organizing a business.

Nonbusiness entities include governmental agencies, foundations, and religious and social groups. Since they are not-for-profit entities, accountants use different account labels to identify their financial transactions and events.

Personal entities are individual persons. You and I are each a personal entity. We can use accounting-type historical records and similar information to describe our resources and the transactions and events that occur.

Figure 12–1   The DAN-D-Acres corporation represents group owners who do business as a single entity. (Courtesy Barbara Doster)

Transactions and Events

Transactions are exchanges of resources or claims *between* entities. For example, you trade some of your money at a store for some cake mix and eggs. Events are recognition of a change in resource cost *within* an entity. For example, you use the cake mix and eggs to make a cake. The cost of the mix and the eggs becomes the cost of the cake resource you now have.

Resources and Claims

The resources of an entity are called assets. The claims are by lenders and/or by owners. As financial historians, accountants want to recognize the claims as being equal to the cost of the resources. Therefore, in the records of accountants, resources must equal claims. Thus,

$$\text{Resources} = \text{Claims}$$
$$\text{or}$$
$$\text{Assets} = \text{Claims}$$
$$\text{or}$$
$$\text{Assets} - \text{Claims} = 0$$

Double Entry Bookkeeping

In a double entry bookkeeping system, accountants maintain the basic accounting equation is:

$$\text{Resources} = \text{Claims}$$

Accountants insist on maintaining the equality of resources (assets) and claims. They have developed a *precise* set of double entry bookkeeping rules. The balance sheet, Figure 12–2, follows these rules. Rules make it easier to identify and track each transaction and event entry to determine what happened.

THE XYZ CORPORATION
Anywhere, U.S.A.

Balance Sheet as of 12-31-90		Assets	Liabilities
Inventory	Books	6735	
	Equipment	1125	
	Supplies	389	
	Accounts Receivable	2189.50	
	Checking Account	2219.83	
	Petty Cash	26.19	
	Total Assets	12684.52	
Accounts Payable	Commissions		1175
	Expenses		216.87
	Paid in Capital and Surplus		11292.65
	Total Liabilities		12684.52

Figure 12–2 The balance sheet represents the idea of double entry bookkeeping—the assets equal liabilities and the books balance.

The amount of assets, and thus the amount of claims, in an entity changes frequently. Accountants recognize two types of changes in an entity's resource and claim amounts. First, the entity can receive money from its owners or pay out financial assets to its owners. Second, the entity may earn a profit (or realize a loss) and thus increase (decrease) its resources and claims.

Later, we will add changes in asset market value to this list. Accountants generally insist on recording only transactions that have actually occurred. Until the asset is sold, the transaction has not occurred. Therefore, accountants generally do not recognize changes in the *market value* of an entity's assets.

Until an asset is sold, accountants recognize only its cost, but the cost of an asset can change. Some assets are completely consumed as a business makes new assets. For example, cake mix and eggs are completely used in the manufacture (baking) of a cake. Other assets are partly used up in the manufacturing process. A corn planter is partly used up in the process of planting each field of corn. Accountants regularly record these events within the business. They collect the costs for the ''cake mix and eggs'' assets and summarize them into an asset account called ''cost of unsold cake.''

Chart of Accounts

Accountants decide how many categories to have to report assets and claims. These categories are called accounts. There are asset-type and claim-type accounts. Later we will present examples of both types and illustrate complete charts of accounts.

Historical transaction and event entries are recorded in the individual accounts. Entries may either increase or decrease their dollar amount. Accountants have developed rules for recording entries.

Journal Entries, Debits, and Credits

Accountants or their bookkeepers record transactions and events in a book of original entry called a *journal*. Each line in the book includes a column for the date, so that each transaction or event can be recorded by the date it occurs. If you know the date an event took place, you can look in the journal to learn more about it. Next to the date, there is space for a short description of the transaction or event. Finally, there are two columns for entering the financial amounts related to each transaction or event. Accountants call the left column the *debit* column and they call the right column the *credit* column.

Accountants put only the date and the description on the first line. They use the following lines, one at a time, to record the individual entries for the accounts that were affected by the transaction or event. Using one line for each account, they write the account name, and in the appropriate column, they write the debit or credit amount by which the account was changed.

Remember the basic accounting equation is ''resources equal claims.'' Another basic accounting relationship is ''debits equal credits.'' For each transaction or event recorded, the *dollar amount of the debit entries must equal the dollar amount of the credit entries*. This means at least one account must be debited when another account is credited.

Recognize the significance of entries made in the journal book of original entry. Before the business begins, there are no entries. Then, as the business begins and continues, all transactions between entities and events within the entity are recorded in this journal. There is an equal dollar amount of debit and credit entries. This means that everything that ever happens, financially speaking, is recorded along with the date it occurs in the entity's journal. It also means that, if each entry is recorded accurately, the various reports we will learn about will contain accurate financial information.

Again, remember the basic accounting equation is "resources equal claims." Using accounting rules, you must now learn to record debit and credit journal entries in order to accurately report what happens to the amount of the entity's resources and claims. Entries may either increase or decrease the dollar amount of money value in each account. It is important for you to remember which type of accounts are increased by a debit entry and which are increased by a credit entry. *Resource accounts are increased by a debit entry, and claim accounts are increased by a credit entry.*

Suppose, for example, you were to start a new business by putting $100 of your money in a bank account for "Ajax Company." As Ajax's accountant, you record this transaction by making the following entries in Ajax's journal:

DATE	DESCRIPTION	DEBITS	CREDITS
Oct. 1	Deposit owner paid-in-capital		
	Debit bank account	$100	
	Credit owner paid-in-capital account		$100

Accountants always put debited dollar amounts in the left column and credited amounts in the right column. You have now recorded the transaction that occurred. What are Ajax's resources and claims? Ajax Company now has resources of $100 located in a bank account. It also has claims against those resources and the claim is by you, the owner of the business. Note that both debits and credits increased by $100 and that resources equal claims.

The last transaction that occurs in a business is the return of any remaining money to the owners. For example, suppose you were to sell all the business material assets and pay all the other claims and there is still $75 left in the Ajax Company bank account. At that point you decide to close the business. In Ajax's journal, you recognize this last transaction as follows:

April 23	Return remaining owner paid-in-capital		
	Credit bank account		$75
	Debit paid-in capital account	$75	

If you were to look at each of the accounts you had created for Ajax, they would now all have "0" balances. You may have recorded debit and credit entries in many accounts during the life of Ajax Company, but you have accounted for all the transactions and events by recording them in the journal. Further, by reading through the journal you could recreate the transaction or event for anyone who might need to know what happened.

Types of Resources

Assets is the name accountants use for material resources. Assets are recorded in the journal at their cost to the entity. They can be acquired directly from the entity's owner as illustrated above by the $100 paid in capital transaction. Assets can also be acquired by purchase from another entity and by creation or "manufacture" within the entity. (Remember the eggs and cake mix purchase and manufacture of the cake.) Finally, assets can be acquired in the form of loans to the entity.

Types of Claims

Accountants recognize two types of claims against the assets of an entity. These are lenders' claims and owners' claims. Lenders' claims are called "debts" or "liabilities." When one entity lends money to another entity, the borrower is said to be "in debt" to the lender.

Owners' claims are called "owners' equity" or "*net worth*." These claims might be said to represent both the first and the last claim to the assets of an entity, but that is not true. A lender's claims to the entity's assets come ahead of the owners' claim. Nevertheless, a business entity could not start without an owner. Therefore, the first historical transactions an accountant will record are related to the start-up of the entity by the owner.

Finally, suppose all the material assets of a business entity are sold. The assets are then distributed to the entities who have claims against the business. The lenders' claims are repaid first. Anything remaining is paid to the owners of the business. Therefore, the last entries recording the financial history of the entity would indicate the amount of assets returned to the owner claimants. As they record transactions and events, accountants are careful that their entries include dollar amounts that assure that the basic equation will be true after all entries are recorded.

Accountants create accounts for classifying and summarizing transactions and events they have recorded. Using summary account titles, the basic accounting equation may be rewritten from:

$$\text{Resources} = \text{Claims}$$
$$\text{to}$$
$$\text{Assets} = \text{Liabilities plus Net Worth}$$

The resources of an entity are claimed by either lenders or owners. We know that when a business is liquidated, the lenders get first claim. Therefore, the equation can be rewritten:

$$\text{Assets} - \text{Liabilities} = \text{Net Worth}$$

Using the accountants' basic equation, we can say that the owners' claims to the entity are what's left after satisfying the lenders' claims. Note the significance of this statement and consider who takes the greatest risk. When you start a business, you use at least some of your resources. You may also borrow some resources. If your business fails to make a profit, the lenders may still be repaid until all the assets are sold. If all the assets are sold and the lenders are not completely repaid, your business is said to be insolvent or "out of business."

If the business succeeds in earning a profit, the owners benefit. The lenders merely are repaid the money borrowed from them plus an interest payment for the amount of time it was borrowed by the entity, whether the entity is an individual, partnership (Figure 12–3), or a corporation.

Figure 12–3   Don Connelly and Bill Hamilton are a partnership entity.

Creation of Paid-in-Capital and Retained Earnings Accounts

To recognize earnings or profits, accountants create different types of "owner equity" or net worth accounts. Two of them are called "paid-in-capital" and "retained earnings" accounts. The claim for the owners money deposited goes in the paid-in-capital account. The claim for earnings realized by the business goes in the retained earnings account.

The accounting equation can now be written as:

$$\text{Assets} - \text{Liabilities} = \text{Paid-in-Capital} + \text{Retained Earnings}$$

In the paid-in-capital account, accountants record the amount of money you or other owners paid into the business entity either to start it or to enable it to continue. If the business is incorporated, you receive corporate stock ownership certificates in exchange for the money you paid in. If the business is not incorporated, accountants merely identify your contribution of money or other material assets as owner paid-in-capital.

At the start-up of a new business, the entity will likely have only assets and paid-in-capital. The assets will be money or other material resources you put into the business. In your own business, you put in the resources and you have an equal dollar claim as paid-in-capital on those resources.

Accountants assume business entities are created in order to earn a profit. Therefore, they might rewrite the basic accounting equation by making profits or earnings the dependent variable:

$$\text{Assets} - \text{Liabilities} - \text{Paid-in-Capital} = $$
$$\text{Retained Earnings}$$

When a business entity is first getting started, it will have assets and paid-in-capital. It may have liabilities. It will not have retained earnings. Therefore, at start-up,

$$\text{Assets} - \text{Liabilities} - \text{Paid-in-Capital} = 0$$

Realization of Earnings

Earnings or profits require the passing of time. Someone must *do* something. In economic terms, someone must create new utility by changing an asset's time, form, place, or possession. This means someone in the business entity must make or market a product or service to another entity. The amount of profit is dependent on what has already happened in the business.

To accountants, "profits are earned" or "earnings are realized" when assets are sold for more money than the cost of those assets used to make and market them. Until an asset is sold, accountants merely collect all the costs of the assets used to make it and keep those costs in a "goods-in-process inventory" asset account.

When an asset is sold for more money than it cost, something almost magic happens in the accountant's history record. First, one asset is decreased by the amount of the costs associated with it to date. Second, another asset is increased and by a larger amount. How can this occur? How can the new asset, say money, be entered into the entity's history record for a larger amount than the cost of the asset that is sold by the entity?

We learned earlier that the amount of the resources must equal the amount of the claim. The answer to the question above is, the accountant has not yet entered all of the claim amount. The accountant will recognize the difference between the sales revenue and the cost as profit or earnings, which as you may remember, is part of the net worth of the owners' claim. Therefore, the retained earnings account is increased whenever earnings are realized.

Conversely, the retained earnings account is decreased whenever an asset is sold for less than it cost. That is the risk a person assumes in owning part of a business entity. Remember again that owners' claims include the paid-in-capital account and the retained earnings account. Therefore, if the retained earnings account is decreased, the owners' claims amount is less.

Consider what happens to the amount of assets when earnings are realized. Remember that asset amounts increase. How will this extra money be used? That is a decision to be made by either the business managers or the board of directors. These groups both report to the business entity owners.

A decision might be made to use the new money within the business, say to buy other assets. If so, when a purchase is made, one asset account will be increased and another decreased by the same amount. The money might be used to repay debts. If so, again one asset account will be decreased and one liability account will be decreased. A decision might also be made to pay the money to the business owners as a dividend. When that is done, the accountant will record a decrease in the cash asset account and a decrease in the retained earnings net worth account. After the dividend is paid, the resources and the claims are smaller but are still equal. Business entity owners, or the board of directors who represent them, need to choose which net worth account to reduce when making a payment to the owners, the paid-in-capital account or the retained earnings account.

Types of Transactions and Events

There are several possible types of account transactions and events. Assume there are three types of summary accounts: asset accounts, liability accounts, and net worth accounts. Categorized by the way the accounts are affected, the types of transactions and events are as follows:

- An asset increase, and an asset decrease
- An asset increase, and a liability increase
- A liability increase, and a liability decrease
- An asset decrease, and a liability decrease
- An asset increase, and a net worth increase
- An asset decrease, and a net worth decrease

Asset Increase, Asset Decrease

Transactions can occur between entities that, as far as accountants are concerned for their records, do not change the financial amount of resources or claims. For example, an entity can exchange a money asset for a material asset such as a machine or some seed. Accountants assume that a market transaction between a willing and able buyer and seller represents an exchange of equal value. Therefore, when an entity buys a new asset, accountants recognize merely an increase in the amount of one type of asset and an equal decrease in the amount of another type of asset.

Accounts also recognize events within an entity whereby one asset account is increased and another asset account is decreased by the same amount. Remember, for example, the recognition of the manufacture of cake mix and eggs into a cake. Two asset accounts were decreased, one asset was increased. Accountants recognize that some assets can be partly consumed in the manufacture of a new product. When you make a cake you use an oven. This oven can be used for years to make many cakes. In a business entity which made cakes, accountants might increase the cost of the cake by part of the cost of the oven. By these entries, called depreciation, the cake asset cost would increase, and the oven asset cost would decrease.

Asset Increase, Liability Increase

Accountants assume the purchase price of an asset is the appropriate evidence of its value. Therefore, when a purchase transaction occurs, they recognize it by recording the transaction price. They record the date of purchase, a physical description of the asset, and the dollar amount of the transaction. Finally, they record the source of the assets used in exchange for the new asset. That is, by the second transaction entry accountants record, they indicate whether or not the entity borrowed assets from others to make the purchase. If assets were borrowed, accountants record this part of the transaction as a debt or account payable by the entity. Accountants also call debts ''liabilities.''

Sometimes an entity will borrow money assets from another entity in a transaction that is separate from a purchase. This type of transaction is called a loan. Accountants recognize that the entity's money assets have increased. They also recognize that the amount of money an entity owes to others has increased. The loan is also a debt.

Liability Increase, Liability Decrease

An entity may have bought an asset from another entity at an earlier date, and agreed to pay for it later. Accountants recorded the initial transaction entries as an asset increase and a liability increase under "accounts payable." Now, a decision has been made by the two entities to enter into a new contract. The debt is being changed from an accounts payable to a loan. Therefore, accountants will recognize the new contract agreement by increasing the liability account called loans payable and decreasing the liability account called accounts payable. Note that neither of these transactions changed the total amount of assets or claims.

Asset Decrease, Liability Decrease

Suppose a debt is repaid with cash. The cash asset account and the debt liability account are decreased equally. Note that resources and claims have both decreased by the same amount.

Asset Increase, Net Worth Increase

Suppose an entity sells a cake asset for more cash than the cake cost to produce. The cash asset account is increased more than the cake asset account is decreased. Therefore, assets are increased. The difference is the profit or earnings. This increase is recorded in the retained earnings net worth account. Resources and claims have both increased by the same amount.

Asset Decrease, Net Worth Decrease

Suppose a cash dividend is declared. The cash asset account is decreased. Assuming the retained earnings account has a positive balance, it is decreased by the amount of the dividend. Resources and claims have both decreased by the same amount.

Trial Balance

You can summarize the journal on any date and determine the total resources and total claims. They will be equal because equal debit and credit entries were recorded for each transaction and event.

When you summarize each account, look at the total of the credit entries and the total of the debit entries. Subtract the smaller amount from the larger amount. The difference is the net debit or credit balance for each account. The total debit entries may equal the total credit entries. In that case, the account has no net effect on the resources and claims, and it is generally not included in reports.

The Balance Sheet: A Point in Time Report

Take the net balances from each account and place them in order with all resource accounts first, and all claim accounts last. Note that most of the resource (asset) accounts have net debit balances and most of the claim accounts have net credit balances.

Now check the arithmetic. Do resources equal claims? Subtract the total of any asset credit balances from the total of the asset debit balances. Also subtract any claim debit balances from the total of the claim credit balances. Is the net debit balance of the resources (assets) equal to the net worth credit balance of the claims? The answer is yes, if you did the work correctly. This answer should be yes on any date you choose to summarize the debit and credit balances in all of the journal accounts.

Accountants have created as their basic report a balance sheet for listing the summary of these individual account balances. Typically, they list resource or asset accounts first. They start with cash or money in bank accounts. Then, they list accounts receivable. These accounts represent the amount of money other entities owe this entity for purchases they have made and not yet paid for. The next part of the asset section of the report includes inventories of various description. These include:

- Raw materials inventory assets that will be used to make goods to be sold.
- Goods-in-process inventory assets that are now being used to make goods to be sold.
- Finished goods inventory assets that have been made or purchased for resale, but not yet sold.

The dollar amount recorded with each type of inventory is the cost of assets purchased and used to date to make each inventory item.

Period Reports

As noted earlier, it is possible to record earnings in the retained earnings account on the date they occur. You will learn to do that by making a new balance sheet report after recording each transaction, including the earnings recognition entries. For example, you might create debit and credit entries for each of the different types of assets, liabilities, and net worth transactions and events described earlier in this chapter.

Several years ago, accountants concluded that most users of their information preferred to see only summary reports describing what happened in a business or other entity. Therefore, most earnings reports are prepared for a one-year period, while some are prepared for a three-month period. These reports are called "profit and loss," or "income statement" reports. An income tax report is a period earnings report.

Period Accounts

Accountants have developed a series of accounts used to summarize information regarding the creation and distribution of earnings over a certain time period. These accounts are labeled "revenue," "cost of goods sold," "expenses," and "earnings."

Revenues. Revenues are sales prices multiplied by the number of units sold in a period of time, usually a year.

Cost of Goods Sold. Cost of goods sold is found by adding the asset costs for all the products sold in a specified time period.

Expenses. Expenses are costs, often for administration and secretarial services, that are more associated with conducting the whole business rather than the manufacture or marketing of specific products. These are measured over a period of time and include some or all of the fuel, electric, phone, property tax, or rent expenses. In addition, the depreciation of factory machinery and equipment is often considered a period expense rather than a part of the cost-of-goods. Machinery repairs are more likely to be included as a part of cost-of-goods since there would be no repairs if no goods were produced.

Earnings. Earnings for a period are found as follows:

Revenue − Cost of goods sold − Expenses = Earnings

Retained Earnings

Retained earnings for a period are found as follows:

Earnings − Income tax − Dividends = Retained Earnings

Types of Accounting

The accountant's record of transactions and events helps people both inside and outside the business better understand what happened. These people can then make better decisions about what to do next. Accounting jobs fall into four categories.

"Bookkeeping" is the routine recording of an entity's financial transactions and events in the entity's journal.

"Financial accounting" includes decisions regarding the classification or creation of the chart of accounts or names for the accounts to be used by the bookkeeper. Financial accounting is concerned primarily with the summary and preparation of reports of transactions and events—reporting revenue, cost of goods, expense and thus earnings for a period of time. By subtracting cost of goods and expenses from sales revenue, accountants estimate earnings or profit for a period of time, usually a year. These reports are created for persons outside the business entity as well as inside managers. Outside persons include owners and persons who are considering becoming owners, as well as tax collectors, who are often trained accountants.

"Cost or Responsibility" accounting is concerned mainly with estimating and accumulating costs for making individual products (goods or services). This information is used by persons inside the business to improve their own performance and that of persons working for them. It is also used in developing future plans for the business. Since products are often being made on a continuing time basis, cost accounting periods are usually less than a year in length.

"Auditing" is a type of accounting used to check on the accuracy of others, mainly financial accountants. "Certified Public Accountants" are experienced accountants who have passed tough exams indicating that they are expert accountants. CPAs may work in any of the specific types of accounting.

SELF-EVALUATION

Balancing Your Books

Multiple Choice: On a separate sheet, write the letter representing the best answer to the question.

1. The journal refers to the
    a. check register of the company.
    b. book the transaction is recorded in first.
    c. specific account credited.
    d. specific account debited.

2. A debit is an entry in the books that
    a. appears in the left column.
    b. appears in the right column.
    c. increases the claims account.
    d. decreases the asset account.

3. Material resources are known as _____ by accountants.
    a. claims
    b. journal entries
    c. account balances
    d. assets

4. The two types of claims the accountant recognizes are called
    a. debits and credits.
    b. journal and accounts.
    c. lenders and owners.
    d. resources and assets.

5. Another name for net worth is
    a. total resources.
    b. owners' equity.
    c. total assets.
    d. total liabilities.

6. Accountants use _____ to summarize financial history.
    a. debits
    b. credits
    c. assets
    d. accounts

7. The formula for net worth can be written as
    a. assets + liabilities = net worth.
    b. resources = claims.
    c. assets − liabilities = net worth.
    d. resources + claims = net worth.

8. The greatest risk in a business venture is taken by
    a. the lender.
    b. the entrepreneur.
    c. the customer.
    d. the employee.

9. Earnings or profits are realized only when
    a. the transaction is properly recorded.
    b. the sale is made and paid for.
    c. the accountant completes the report.
    d. the sale becomes paid-in-capital.

10. Depreciation is the accountant's way to account for _____ an item such as a delivery truck in the conduct of business.
    a. using up
    b. repairing
    c. replacing
    d. purchasing

11. This depreciation is recognized by
    a. increasing an asset cost account and decreasing another asset account.
    b. increasing a liability account.
    c. decreasing a liability account.
    d. decreasing the owners' equity account.

12. A trial balance refers to a
    a. summary of asset accounts.
    b. summary of liability accounts.
    c. summary of total assets and liability.
    d. statement listing only the owners' net worth.

13. A balance sheet is
    a. a point in time report.
    b. a listing of account debit balances.
    c. a listing of account credit balances.
    d. evidence of proper accounting procedures.

14. On a balance sheet
    a. assets will exceed liabilities.
    b. liabilities will exceed assets.
    c. net worth will be the largest amount.
    d. assets or resources will equal claims or liabilities.

15. Raw materials are always listed in
    a. asset accounts.
    b. liability accounts.
    c. goods-in-process accounts.
    d. finished goods accounts.

16. Cost of goods sold is
    a. equal to selling price.
    b. equal to raw materials cost.
    c. equal to the costs directly attributed to the goods to be sold.
    d. the asset costs for all the products sold in a time period.

17. Bookkeeping is
    a. the assigning of costs to individual products.
    b. the recording of financial transactions.
    c. called auditing.
    d. the application of accounting practices.

18. A credit sale of $100.00 worth of fertilizer would be entered
    a. in the journal.
    b. in the journal and in accounts receivable.
    c. in the journal, accounts receivable, and in a fertilizer account.
    d. in the accounts receivable and the asset account.
19. The basic information on the sales ticket must include
    a. the date, the name of the customer, description of the item sold, the amount of the sale, and the type of payment (cash, check, charge, or credit card).
    b. the name, description, and type of payment.
    c. the date, customer, seller, and terms of payment.
    d. the amount of the sale and description if it is a cash sale.
20. Accountants are
    a. bookkeepers.
    b. financial historians.
    c. CPAs.
    d. managers.

STUDENT ACTIVITIES

1. Keep a similar set of records to reflect a period of time for your proposed business plan. (Or use Kermit's journal.)
2. Develop a period earnings report based on your first year's projections. (Or use Kermit's journal.)
3. Obtain a business journal and post the proper double entry items for typical transactions. Kermit's journal is an example.

KERMIT'S SWEET CORN JOURNAL

DATE	ITEM	DEBIT	CREDIT
March 1	To start Kermit's checking account balance		
	Debit checking account	90.43	
	Credit paid in capital account		90.43
March 15	Rental of lots from Lyle Graham		
	Debit goods in process	75.00	
	Credit bank account		75.00
April 15	Loan from Uncle John		
	Debit checking account	70.00	
	Credit loan account		70.00
April 16	Exeter Feed and Grain—fertilizer		
	Debit goods in process	37.75	
	Credit checking account		37.75

April 26	Spoo Hardware—sweet corn seed		
	Debit goods in process	22.50	
	Credit checking account		22.50
May 7	Stone's General Store—hoe		
	Debit equipment	6.83	
	Credit checking account		6.83
July 5	Exeter Leader—advertising		
	Debit goods in process	5.25	
	Credit checking account		5.25
July 7	Sale of sweet corn		
	Debit checking account	11.00	
	Credit goods in process		11.00
July 8	Sale of sweet corn		
	Debit checking account	15.00	
	Credit goods in process		15.00
July 9–31	Sale of sweet corn		
	Debit checking account	182.50	
	Credit goods in process		182.50
July 31	Repayment of loan to Uncle John		
	Credit checking account		72.50
	Debit loans	70.00	
	Debit goods in process	2.50	
August 27	Depreciation of hoe		
	Debit goods in process	2.28	
	Credit equipment account		2.28
August 1–27	Sale of sweet corn		
	Debit checking account	382.75	
	Credit goods in process		4.78
	Credit retained earnings account		377.97

CHAPTER 13

Decisions, Decisions. . . How Do I Make Them During The Start-up Period?

OBJECTIVE

To examine a managerial decision-making model.

COMPETENCIES TO BE DEVELOPED

After completing this chapter you will be able to:

1. Identify the major items needed in a financial plan.
2. Identify the potential sales to use in making cash flow projections.
3. Make realistic estimates of cash flow projections.
4. Identify appropriate target customers.
5. Develop a decision-making procedure for the start-up period.

TERMS TO KNOW

Canvass

Compassion

Family living expenses

Listed

Pest control

Sales supervisor

Service club

Start-up

Cash flow projection

Considering

Fixed cost

Major

Recapitulate

Secure

Spouse

Termination

INTRODUCTION

The next chapter deals with the use of records and reports in managerial decision making. In this chapter we want to provide help in making *start-up* decisions, an area in which information is often hard to obtain.

We will discuss many elementary problems that might be unforeseen. When decisions need to be made on short notice, they have greater chances of being wrong and costly. If you plan for decisions, you will be better prepared to make them.

In the start-up phase of a business enterprise one must do a careful cash flow analysis that reflects not only the needs of the business but also the living expenses for the entrepreneur and his or her family. We will present two case studies to illustrate these concepts and other decisions necessary to get a business started.

CASE STUDIES

The Case of the Planned Parting

You met Ted and Mary Ruth in Chapter 2, but we will *recapitulate* their story to illustrate many of the start-up decisions that are typical with a new enterprise. We will retell their story in a later chapter to illustrate other points closely allied to these.

Ted worked for a *major pest control* company and had done well in sales. He had done so well that he earned a promotion. The promotion meant he would have to move to another city, and he had no choice but to accept the promotion and move if he wanted to stay with the company.

Ted and Mary Ruth did not want to move. Ted also did not want to move into the *sales supervisory* position that was to open up four months down the road. Mary Ruth had a nursing job and thought she could help support their basic needs as Ted developed a new pest control business. They spent many evenings *considering* possibilities for a start-up business.

Ted had five years of experience working in the pest control business. This gave him a good working knowledge of what would be required to start such a business. Accordingly, Ted went to the banker and applied for a loan, but the first request was turned down. They were still convinced Ted could launch a successful pest control business, but they realized that their financial plan was not adequate. They went to work on a more accurate *cash flow projection* to present to the bank.

Ted and Mary Ruth decided that with her nursing income, they could cover the mortgage payments, *family living expenses* (at a reduced level), and about 80% of the first two years' expenses. They would use up much of their savings for start-up expenses and for hiring someone to answer the telephone and do the clerical work while Ted answered pest control calls.

To keep expenses low Ted planned to put his office in the basement and hire only one employee until sales expanded to support others. Then Ted could devote additional time to sales activities—the part of the operation he preferred. The decision for a home-based business was only part of Ted's location problem. The other was a suitable location for *secure* storage of pest control chemicals and applicators.

Ted and Mary Ruth next made the decision not to let the pest control company know of their plans because Ted would probably be dismissed at once. Ted needed the steady income while the necessary planning for his venture was completed.

Ted and Mary Ruth also faced the problem of how to obtain customers for their new business. They had missed the advertising deadline for the yellow pages so another advertising plan was needed. One strategy was to contact the customers Ted had worked with successfully for five years. He could offer to continue the work as a local private company instead of representing the national company.

After considering a number of advertising plans, they decided to budget a certain percentage of each sale for promotion after the initial advertising budget was expended. This would be a chance to tie advertising to revenues and would simplify cash flow projections. They recognized, however, that the budget might be low when the need was greatest.

We will return to Ted and Mary Ruth's situation to look at their cash flow projections. First, let's look at another case.

The Case of Jim's Jump

Jim had taught successfully in the local school system for eight years. He was now looking for a new challenge and an opportunity to improve the family income so that college would be a possibility for his two daughters. Jim's wife, Jill, was four months pregnant with their third child. They realized that in the short run they might have to tighten their belts and change their life style. Taking stock, they listed their strengths to help with the decision-making process.

- Jim was a popular teacher.
- Jim was a five-year member of the local Kiwanis club and had been active in working for community improvement.
- Jim had an uncle who was an independent insurance agent in a neighboring state.
- Their good friends, Don and Janet, had started their own insurance agency two years earlier and were showing a profit. Don had agreed to help Jim learn the insurance business as they would not be in head-to-head competition.
- Insurance and sales both interested Jim.
- Jill was a former teacher and had extensive office experience after high school and during college.
- Jill could manage the home office with some temporary help when the baby was born.
- They had some savings that could help them through the start-up phase.
- Jill's pregnancy expenses were covered through the school's group policy until the baby was born.
- Without the threat of summary dismissal, Jim notified his principal and superintendent as soon as the decision was made. They set the changeover for the semester break. (Jim got his principal and superintendent as two of his first insurance customers.)

Family living costs were one of their first concerns. Their annual budget is shown in Figure 13-1. The first column shows the actual expenditures for 1990. (It is most helpful to show the figures as a future projection, hence 1991 estimates.) The estimated expenses are charted for the following year.

Note that child care is higher in July, for at this time they did not know Jill's mother would come in to help through the birth period. Note also that clothing costs are up in August as the two children are outfitted for kindergarten and second grade. Education is also up in August when school fees are paid. Gifts show an increase in December as they do not plan to change their Christmas giving.

Note that the projected total expenditures are less than in the previous year. Jim and Jill recognize that short-term sacrifices are necessary in deciding how to use what they now have to get what they now want most—in this case successfully starting their own business.

J and J Insurance had some start-up decisions to make and some facts to gather as they planned for the mid-January change in occupations.

- How could they obtain independent agency status for several lines of insurance?
- How could they gain recognition as a viable business?
- Would city ordinances permit an advertising sign in their front yard and if so, what size?
- How could they get a telephone listing? Had the new telephone directory gone to press?
- Would a home office be adequate in the start-up stage?
- Could Jim find training sessions to help him learn the insurance business?

These and many other questions were answered during the start-up phase of the business.

What the Cases Tell Us

- When you work for a national company you may have no choice about where you work.
- Many companies have very little *compassion* for employees' families.
- Company best interest may not be personal best interest.
- Two-job families may find entrepreneurship easier than one-job families.
- Family living expenses as well as business cash flow must be clearly planned.
- Families will often need to sacrifice in order to start an enterprise.
- How employees treat planned *terminations* varies greatly.
- Experience is valuable in projecting realistic costs and returns.
- *Service club* membership helps businesspeople know other businesspeople.
- All business people need to work to improve their communities.
- Relatives and friends are important information and support people.
- A beginning entrepreneur needs an active interest in the chosen field.
- A supportive *spouse* is a must in a successful business start-up.
- Projected cash flow and expenses should be realistic and conservative.
- Attention to city ordinances and other legal matters is essential.

FAMILY BUDGET PROJECTION

FAMILY LIVING EXPENDITURES	1990 TOTALS	JAN	FEB	MAR	APR	MAY	JUNE	JULY	AUG	SEPT	OCT	NOV	DEC	1991 TOTALS
1 Food	6724	525	525	525	515	495	490	500	510	525	530	535	600	6275
2 Clothing	2166	75	75	75	75	80	150	80	225	75	75	75	150	1210
3 House payments	4488	374	374	374	374	374	374	374	374	374	374	374	374	4488
4 Home repairs	59							200						200
5 Furnishing	369					200								200
6 Utilities & fuel	1485	125	120	110	100	90	90	95	110	100	120	135	190	1485
7 Education	350	40	35						125					200
8 Recreation	794	20	20	20	20	20	20	20	20	20	20	20	20	240
9 Transportation	2215	85	85	85	90	100	80	90	80	80	80	80	80	1005
10 Health	480	40	40	40	40	40	40	40	40	40	40	40	40	480
11 Personal care	250	20	20	20	20	20	20	20	20	20	20	20	20	240
12 Health insurance	180	138	138	138	138	138	138	138	138	138	138	138	138	1656
13 Life & other insurance	2437	51	51	51	376	51	51	51	51	51	376	51	326	1537
14 Child care	525	20	20	20	20	20	60	200	20	20	20	20	20	460
15 Gifts & donations	1350	25	25	25	25	25	25	25	25	25	25	25	175	450
16 Savings	1050													
17 Other expenditures	431	25	25	25	25	25	25	25	25	25	25	25	25	300
Total	25400	1613	1603	1588	1928	1628	1512	1858	1743	1493	1843	1538	2060	16086

Figure 13–1 Jim and Jill's family living costs. (Courtesy The Three Entrepreneurs)

- In the start-up phase of a business many minor decisions will also be necessary. Among those later made by the couples were
  — design and printing of letterheads
  — bookkeeping system selection
  — licenses and permits needed
  — setting up a separate business checking account

DETAILED CASH FLOW ANALYSIS

Looking More Closely at Some Start-Up Decisions

We examined the family living expenses of Jim and Jill. In this section let's look at the cash flow analysis of Ted and Mary Ruth's start-up period. Like many others, they will make sacrifices at this time, Figure 13–2.

Figure 13–2   Janet and Don ate hamburger instead of steak while getting their business started. (Courtesy Janet Connelly.)

Their costs will include:

1. The secretarial employee budgeted at $7.50 per hour, including insurance, social security tax, and so on.
2. Advertising budgeted at $400 plus 5% of sales.

3. Equipment

Truck—owned value	$5,995
company name to be painted	125
Applicators	600
Protective clothing, respirators and filters	800
Office equipment	
Desk on hand	
Secretary's work station	1,100
Personal computer, printer, word processing software	1,200

4. Supplies

Office	400
Bait stations—inventory replaced or billed to customers	150

5. Operating Expenses

Truck—gas, oil, maintenance, insurance, depreciation, license, and so on.	$150/month
Telephone service charges	$35/month
Utilities—share of cost	$100/month
Chemical purchase at 30% of sales	
Applicators	
Respirators	
Insurance	

As we look at Ted and Mary Ruth's cash flow projections we note several items of special interest. First, after the initial start-up expenses the highest *fixed cost* is the secretary's wages and benefits. Second, the initial advertising budget is underwritten until sales provide the needed percentage to provide a level funding. Third, variable costs for chemicals and supplies are tied to sales projections. Fourth, the break-even point is projected to occur in the ninth month, which is not unrealistic according to Ted's experience.

Ted planned to *canvass* his former customers as soon as he opened the new business and has made conservative estimates of those who would change to a local operator. Ted thought this approach to the estimate best after having been denied the loan initially.

Ted and Mary Ruth estimated their family living expenses and projected using their savings to cover the remaining expenses the first year. This estimate was made at $4,000, leaving $6,000 available to help with start-up costs. The net start-up cash needed was approximately $4,500 with $1,200 more estimated for January and $900 for February.

Accordingly they asked for a line of credit of $12,000, with the option of renewing it for a second year. At this point projected cash flow would allow regular repayments of $500 per month in February and $1,000 by June increasing to $1,500 in August. This would put the business on sound footing with only a small balance carried into the third year. Figure 13–3 details their cash flow projections.

ITEMS	START-UP TIME	JAN	FEB	MAR	APR	MAY	JUNE	JULY	AUG	SEPT	OCT	NOV	DEC	TOTAL FOR YEAR
EXPENSES														
1 Secretarial support		920	920	920	920	920	920	1000	1000	1000	1000	1000	1000	11520
2 Work station	1100													1100
3 Typewriter	250													250
4 Used PC and printer	1200													1200
5 Truck (owned)	(5995)													(5995)
6 Operating expenses	310	125	125	125	125	125	125	125	125	125	125	125	125	1800
7 Personal protective clothing	600							100						700
8 Applicators	800							100						900
9 Warehouse rent	190	190	190	190	190	190	190	190	190	190	190	190	190	2470
10 Supplies for resale	2000	100	100	100	100	125	125	125	125	150	150	150	150	2400
11 Printing	250				50			50						350
12 Office supplies	250								50				50	350
13 Advertising	250	100	100	100	100	125	140	140	140	150	150	200	200	1895
14 Telephone	250	50	50	50	50	50	50	50	50	50	50	50	50	850
15 Utilities (share)		100	100	100	100	100	100	100	100	100	100	100	100	1200
16 Other expenses	1000	50	50	50	50	50	50	50	50	50	50	50	50	1600
17 Chemical purchases	2000	210	300	335	350	390	400	500	550	600	635	700	800	6770
Total Cash Needed	16490	1930	1935	1970	2035	2075	2100	2570	2380	2315	2350	2565	2715	40430
SALES														
1 Service		700	900	1000	1150	1300	1450	1600	1800	2000	2250	2500	2750	19400
2 Supplies		150	150	175	200	225	250	250	250	275	275	300	300	2800
TOTAL SALES		850	1050	1175	1350	1525	1700	1850	2050	2275	2525	2800	3050	22200
Cash Available	11995	-1200	-885	-800	-685	-550	-400	-720	-330	40	175	300	250	7190
Loan Proceeds Needed	4495	1200	885	800	685	550	400	720	330					10065
Savings Utilized	5000													5000
Cash Available for Debt Repayment											125	300	250	675

Figure 13-3 Ted and Mary Ruth's cash flow projection. (Courtesy The Three Entrepreneurs)

SUMMARY

Looking at these two start-up businesses, one with an experienced operator and one without, we have studied decision making. The examples illustrate the teamwork between husband and wife when a new business is being born. It is worthy of note that these decisions involved

When to start

Where to start

What funds are needed

When funds will be needed

How to attract business

How the business will affect the family

Sales projections

Cash flow analysis

Insurance, both health and business

It is important to consider that business decisions at this stage are intertwined with family matters. Again the DAN-D-Acres business reflects how start-up decisions are made and continued to be made as the business grows, Figures 13–4 and 13–5.

Figure 13–4 Doing many things by hand is common in the start-up phase. (Courtesy Barbara Doster)

Figure 13–5   **When the business is established, money is available to purchase equipment. (Courtesy Barbara Doster)**

SELF-EVALUATION

Decisions During the Start-Up Quiz

True-False: On a separate sheet, write T for true statements and F for false statements.

1. Family living costs need not enter into financial planning.
2. Financial plans should be as realistic as possible.
3. It would be better to err on the optimistic side.
4. An advertising budget should be tied to sales in the start-up period.
5. In starting a business, an individual should solicit helpful advice from people with experience.
6. Cash flow projections are a useful planning tool.
7. Experience in the field will help an entrepreneur gain financial support.
8. Starting a business may mean short-term sacrifices.
9. A line of credit refers to the total of money borrowed.
10. The business plan not only should tell how much money will be needed, but when it will be needed.
11. Two-job families may find entrepreneurship easier than one-job families.
12. An entrepreneur is not concerned with community improvement.

13. A good manager does not need to have a high interest in the product area in order to do well.
14. A supportive wife or husband is a great asset to the beginning entrepreneur.
15. National companies will give you a choice of where you want to work.

STUDENT ACTIVITIES

1. Develop a list of decisions you will need to make in starting your own business.
2. Develop a monthly cash projection for the first two operational years of your business.

CHAPTER 14

In Decision Making, How Do Records Help Me Project Cash Flow?

OBJECTIVE

To explore how records and reports can help in making management decisions.

COMPETENCIES TO BE DEVELOPED

After completing this chapter you will be able to:

1. Identify the component parts of a balance sheet.
2. Identify the component parts of a profit and loss statement.
3. Calculate the ratio of owners' claims to others' claims from a balance sheet.
4. Calculate the ratio of current assets to current liabilities given the year's balance sheet and profit and loss statements.
5. Calculate the degree of solvency given the year's balance sheet and the profit and loss statements.
6. Describe the cash flow projection for a business given a year's reports.

TERMS TO KNOW

Budgeting	Debt service
Firsthand	Floor samples
Income statement	Lines
Residual	Solvency

INTRODUCTION

Records and reports are valuable decision-making tools. Unless we know where we have been, we will have trouble deciding where to go. If we need to decide what to do with an enterprise or part of an enterprise, we need to know whether or not that portion of the business is profitable.

In this chapter we will highlight how records and reports can aid in informed decision making. We will look at the period reports and suggest some other reports one can use in *budgeting* and decision making. The reports an entrepreneur wants to study include:

- sales,
- accounts payable,
- accounts receivable,
- income tax,
- profit and loss,
- balance sheets, and
- individual accounts.

Among the reasons for studying these kinds of reports are to:

- increase company profitability.
- promote the company image.
- increase the company share of a certain market.

In a sole proprietorship it is likely that the owner-operator can have the overall picture from *firsthand* experience. In other, larger businesses it is less likely the owner(s) will have such a complete picture, hence the need for good records and reports.

CASE STUDY

The Case of Local Approval

At the end of his first year in business, Phil decided it was time to look at the *lines* he had added to the Exeter Feed and Grain business. A local manufacturer of wooden feeders had approached him to take on their line of feed bunks and hog feeders. Phil appreciated the local vote of confidence in his new business but wasn't sure he could afford the investment of limited capital needed to carry the required inventory and *floor samples*.

Since Phil was still the bookkeeper for the business, he knew which accounts he needed to examine in order to make a valid decision. Phil examined the total of monies invested—the returns or sales—and how each of the lines contributed to the business.

One of the better returns was the all-metal line of hog equipment. This was first added through the discount order procedure started early in the year and had now grown to an inventory status. Phil's conclusion was that a limited inventory of wooden equipment would be a complementary line.

What the Case Tells Us

- Records and reports are valuable decision-making tools.
- Records need to be complete.
- Records must be accurate.
- Records need to be organized so one can easily gather needed information.
- Other factors besides records are needed in decision making.

RECORDS AND REPORTS IN DECISION MAKING

Gathering Needed Financial Information

The typical data needed for managerial decision making will include basic financial reports such as

- Balance sheets,
- *Income statements*,
- Sales reports,
- Individual account records, and
- Cash flow analysis.

Balance Sheets.   Several chapters have discussed the company balance sheet. Let's examine this information again through our decision-making glasses. Figure 14–1 is a theoretical balance sheet for Willy's Widgets Company.

ASSETS

Current assets	
Cash (checking account)	7,657.92
Accounts receivable	11,645.00
Inventory	
Finished widgets	14,000.00
Raw materials	2,580.00
Prepaid expenses	480.00
Plant assets	
Land	25,000.00
Buildings and equipment	114,000.00
Total assets	175,362.92

LIABILITIES

Current liabilities	
Accounts payable	8,122.50
Operating loans	15,000.00
Mortgage on property	35,000.00
Owners equity	90,000.00
Retained earnings	27,240.42
Total liabilities	175,362.92

Figure 14–1   Willy's Widgets balance sheet. (Courtesy The Three Entrepreneurs)

The balance sheet shows that the owners have $117,240.42 of the business's assets, while there are claims on the other $58,122.50 of the company's total assets of $175,362.92. In other words, the owners have $3.03 in assets for every $1.00 in claims by others. Most business analysts consider this a positive financial position—they would not consider the business in immediate danger. If liabilities other than the owners' equity approach 1 to 1 the danger flags are raised.

As stated earlier, the balance sheet is a status report at a single point in time. This information can only be compared to other past points-of-time reports. Decision making is based on what will likely happen in the future as we try different recipes of how we might use what we now have to get what we now want most.

The Income or Profit and Loss Statements.   Again we look at Willy's Widgets Company's yearly reports for 1989 and 1990 shown in Figure 14–2. These two statements show the performance but do not reveal the details of how the business functioned during the year. One quarter may be more active than another. From the balance sheet we see that the inventory at the end of the year is minimal, but we don't know what happened during the year. In a sole proprietorship, the owner or manager will likely know these facts through involvement in the business. In a partnership there is less chance of the knowledge being firsthand, and in a corporation it is unlikely the knowledge would be firsthand. In the latter case, a different period-of-time report would be useful to provide the desired information. In such cases a quarterly report, as created by major corporations, would be helpful.

	1989	1990
Sales	58,000	210,000
Cost of goods sold	86,440	108,460
Gross profit	71,560	101,540
Operating costs	33,180	39,900
Interest charges	4,880	5,120
Income before taxes	33,500	56,520
Taxes	7,035	12,434
Net income after taxes	26,465	44,086

Figure 14–2   Willy's Widgets profit and loss statement. (Courtesy The Three Entrepreneurs)

Sales Reports.   The owner or manager needs to have current sales figures as well as past sales records in order to maintain adequate inventory and be able to schedule services. In a production business, raw materials need to be ordered on time in order to renew or maintain an adequate inventory cushion and to facilitate timely shipment of finished goods. Workers will need to be scheduled to maintain production and the cycle continues. In Figure 14–3, a worker assembles his product. Sales reports are vital to these other facets of the business.

Records and reports can be used to predict future growth, plan production, and set sales goals. The sales report will record sales volume in both units and dollars. Each separate item or category of items will be reported as needed. The facts will be gathered from sales slips, and daily or monthly business summary totals. By filing these sales reports quarterly, semiannually, and annually, totals can be amassed. These can then be used for next year's business cycle.

Figure 14–3   Mark is manufacturing fishing lures for sale. (Courtesy Janet Rausch)

The further removed the owners are from firsthand knowledge, the greater their need for up-to-date sales figures. Decisions are worth more when made at the proper time than they are when made at the end of the week, the month, the quarter, or the year. Timely decisions are made when pertinent information is available. Testing your product can aid in decision making, Figure 14–4.

Individual Accounts.   A balance sheet does not show any accounts that have no balance. However, the manager needs to examine many of these accounts as decisions are made. In the case of Exeter Feed and Grain, Phil looked at the Energy Feeder Company account, which detailed the sales of all-metal hog feeders and waterers. The number of sales and their contribution to earnings were useful in making the decision to handle the Karmel wooden feeders.

This information required a different set of data than that provided by period reports and the balance sheet. Getting at such information is easier when using double entry versus single

Figure 14–4 Mark takes time out to test his product. (Courtesy Janet Rausch)

entry bookkeeping. The single entry system would have required looking over each journal entry and selecting the ones wanted. In the double entry system, the information was consolidated into one account.

Cash Flow Analysis.   A review of cash flow for the first year provided Phil with information concerning the availability of cash needed to add the new line. In this case, January was an excellent cash position month because many customers reported taxes on the cash basis and paid off their accounts during December.

Accounts Payable.   In a single entry system the records will likely be found in a desk drawer or folder, while in a double entry system they will be detailed in the books of the company. We want to mention them here as decisions must constantly be made as to when to "put the check in the mail." An entrepreneur does not want to be put in suppliers' books as "slow to pay"; the business might wind up on a slow shipping list. By keeping a close watch on payables, many discounts can be used to lower costs and thus aid profitability.

Accounts Receivable.   Detailed individual accounts are available to show the status of accounts due to the business. Many decisions on credit extended by the business develop from studying these accounts. Often, managers are rated on how these accounts are managed.

Analysis of Financial Reports

After we have gathered the financial data available for our business, we will look at some indicators of business strengths or weaknesses. One of these indicators looks at how your current assets compare to your current liabilities and principal payments on long-term liabilities. In the case of Willy's Widgets this comparison would reveal the following facts

Current assets	$36,362.92
Current liabilities	23,122.50
Year's principal on mortgage	2,000.00

Note that only the principal is included. Interest is already charged against operating expenses.

We can see that Willy's Widgets has current assets of $1.45 for each $1.00 owed as current liabilities. In most financial circles, any time your liquid (easily converted to cash) assets fall below $1.50 for each $1.00 of current liabilities, caution is advised. If this relationship approaches $1.00 to $1.00, there is no cushion for adversity such as a slump in sales.

A second measure we should look at in considering the reports from Willy's Widgets is the degree of *solvency.* This concept relates to the soundness of the business over the longer term. The basic figure for this measure was discussed after the balance sheet was listed earlier. In this case, Willy's Widgets has assets of $3.03 for each $1.00 of debt and is very solvent over the long run.

Lenders will judge this relationship as favorable when operating loans are requested. One of the management decisions triggered by this analysis could be to work on accounts receivable in order to lower the need for operating loans. Inventory management indicates that the completed goods inventory contains less than a 30 day supply and raw materials are probably critically low. Unless shipments are regular and dependable, further attention may be required to increase this inventory. This again emphasizes the single dimension of a point-in-time report. The raw materials mentioned could have been delivered the next day.

When we look at individual account records and the cash flow information, we are in a position to know our total operating receipts and total operating expenses. We can compare the timing of the receipts to the payment of expenses. The *residual* at the end of the year can be seen as a cushion against adversity.

Look at our information on Willy's Widgets. We find the company has $23,122 of current liabilities, a $2,000 principal payment, and $5,120 in interest payments, totalling $30,242 in *debt service* in 1990. When we compare this to the annual cash income of $210,000, it represents 14.4% total cash revenues. If this ratio exceeds 25%, observe caution. When the percentage is less than 15%, it is a sign of security. In this case Willy's Widgets should be cautious in taking on larger loans.

Projecting Next Year's Cash Flow

The cash flow projection for Willy's Widgets is presented in an abbreviated form (showing each source of expenditures contributing to cost of goods sold). It is based on the year's sales goals. Figure 14–5 shows that Willy's Widgets can pay off the operating loan through retained earnings as accounts receivable and inventory total $43,500 at the end of the year, while they are $25,645 on January 1st.

CASH FLOW PROJECTION

Revenues

Beginning cash balance	$7,657.92
Beginning inventory of widgets	14,000.00
Accounts receivable	11,645.00
Projected sales	250,000.00
	283,302.92

Expenses

Cost of goods sold	129,070.00
Operating costs	47,082.00
Interest charges	4,800.00
Taxes	13,500.00
Contingency fund	5,000.00
	199,452.00
End of year inventory	23,500.00
End of year accounts receivable	20,000.00
Cash on hand	40,350.00

Figure 14–5 Cash flow projection for 1991. (Courtesy The Three Entrepreneurs)

If the business is not subject to wide seasonal variations, it will be able to pay off the operating loans through projected cash flow. The only time cash flow appears to be a problem is at the beginning of the year when raw materials are purchased. This is evident when we see an inventory of only $2,500 on hand. Sales, if they are uniform, will soon catch up. January's share of materials would be about $10,600 while cash was less than accounts payable. With January sales of $20,000 the cash flow should ease in February or March. Similar projections could be made for any month or quarter based on previous records and projected goals.

SELF-EVALUATION

Records and Decision Quiz

Use this example in answering the following questions.
Joe's Diner and Catering Service has the following year end assets and liabilities.

Current Assets

Checking account balance	$1,672
Accounts receivable	6,289
Inventory	2,216
Prepaid expenses	614

Assets
Diner and fixtures	116,490
Total assets	127,281

Current Liabilities
Accounts payable	4,622
Operating loans	2,500

Liabilities
Mortgage on property	21,000
Owners equity	50,000
Retained earnings	49,159
Total liabilities	127,281

Principal payment of $3,000 per year

Interest payment of $330 per year

Multiple choice: On a separate sheet, write the answer to the question.

1. The ratio of owners assets to claims of others is approximately
   a. $3.75 to $1.00 of others claims.
   b. $3.50 : $1.00.
   c. $2.75 : $1.00.
   d. $2.50 : $1.00.

2. The current assets:current liabilities ratio shows the business has
   a. $2.00 of current assets for each $1.00 of current liabilities.
   b. $1.75 : $1.00.
   c. $1.50 : $1.00.
   d. $1.25 : $1.00.

3. The degree of solvency in the long run is illustrated by question
   a. number 1.
   b. number 2.
   c. neither question.
   d. both questions.

4. Joe's Diner has a total debt service of $5,380 for the current year, if his cash income totals $88,000. What is his debt service ratio?
   a. 5%.
   b. 6%.
   c. 7%.
   d. 8%.

5. Using the ratios determined in the earlier questions, Joe could be _____ about taking on more current debt.
   a. conservative
   b. very conservative
   c. mildly cautious
   d. not worried

6. If Joe wants to expand his facilities, he will likely be able to convince the banker with
   a. little difficulty.
   b. severe difficulty.
   c. ratio reports.
   d. sales projections.

7. In addition to the profit and loss statements and the company balance sheet, the manager needs to be concerned mainly with the
   a. accounts payable.
   b. accounts receivable.
   c. ratio reports.
   d. sales projections.

STUDENT ACTIVITIES

1. Based on your business plans, calculate each of the indexes described.
2. Obtain a local or area published balance sheet and compute these indexes.
3. Work up hypothetical examples of sound and unsound businesses and give suggested plans of action for the managers or owners.
4. Listen to a local banker discuss balance sheets and profit and loss statements.

CHAPTER 15

How Can I Manage Customer Credit and Collections?

OBJECTIVE

To identify how credit affects a business.

COMPETENCIES TO BE DEVELOPED

After completing this chapter you will be able to:
1. Identify strengths and weaknesses of a credit policy.
2. Identify types of credit.
3. Calculate the true cost of credit.

TERMS TO KNOW

Accounts payable	Accounts receivable
Acrimony	Adamant
Bad debt	Chattel
Confrontation	Credit
Credit limit	Credit policy
Monthly statement	Recourse

INTRODUCTION

Credit is a sales tool with advantages and disadvantages. Careful management is needed when credit is used. In this chapter we look at these advantages and disadvantages, different types of credit, credit applications, collections, policy and the use of credit by a business.

CASE STUDY

The Case of the Oldest Debt

Phil, partner in Exeter Feed and Grain, had just returned from a conference with Cecil, his banker, and wished his partners were in town so he could get their advice. He needed another truckload of feed but he was at the *credit limit* set by the feed mill.

In the middle of the summer, feed demands were high as the swine and beef cattle neared market weights. Farmers ordered feed with the time-honored promise, "I'll pay when I sell my hogs (or cattle)." The accounts were basically good because the customers subscribed to the oath, "My word is my bond." However, Phil had an immediate problem. His business assets were excellent in the long run, but cash flow was a problem in the short run.

How was Phil going to raise the cash needed to buy another load of feed? He decided his only *recourse* was to call on customers to pay their past due bills. In beginning this process, he decided he should collect the oldest debt first.

The oldest debt was for a truckload of shelled corn purchased 18 months before by Joe Hearty's son, Ned. Repeated *monthly statements* had been ignored so Phil telephoned the farm and made an appointment to discuss "an important matter." The appointment was set, and Phil drove 27 miles to the farm, which was in an adjoining county. Phil was greeted by Ned's mother, his father, and Ned himself. In the background was Ned's new wife, Lucy, a former employee of Phil's bank. Phil waved to Lucy, then asked if there was a problem with the bill. Mother and son were *adamant* that they had purchased no corn from Exeter Feed and Grain. The father said nothing.

Phil brought out the original sales ticket. After heated verbal abuse by the mother, which Phil ignored, she said, "Give him the money! But if I ever see him on this place again, I'll use the double-barrelled shotgun!" Taking the check, Phil returned to Exeter. A load of grain needed to be delivered to a farm where the portable elevator could not be used. Since Phil thought it best to stay away from people until his anger subsided, he let an employee tend shop and he shoveled the 240 bushels of grain.

The next morning Joe Hearty walked in the door at Exeter Grain. He was an honest man and had come to apologize for his family. He explained that after Phil left, Lucy had charged over to the three at the gate, the scene of the *confrontation*. She cried out, "I never thought I'd be anything but proud of my new name, but today I am deeply ashamed. If Phil says you bought the grain, then you owe him the money. He is a fine person!" After a time, Ned admitted he had a grudge against one of the partners and had decided to cheat them out of the load of corn. He had told his father that the grain had been purchased in Stuart instead.

What Do We Learn from Phil's Rough Day?

- Exeter Feed and Grain needed a better *credit policy*. If a person's word is good, he or she won't mind signing up for *credit*.
- The credit should have been followed up sooner. After eighteen months a *bad debt* should be taken as a write-off.

- Phil was wise to state the facts and not argue about details or engage in *acrimony* (verbal name calling and argumentation).
- Joe Hearty felt compelled to let Phil know how the misunderstanding occurred. Had Phil's temper exploded, Joe probably would not have driven 27 miles to apologize.
- Oldest debts should be cleared if possible before asking for payment on more current accounts. Too much nagging may drive away good customers who don't think they can rely on the agreements made at the time of settlement.
- Most retail businesses should not be in the lending business. Commercial sources should handle credit.
- Credit increases the cost of sales: there is postage for billing, interest on capital, and bookkeeping time involved.
- Credit may alienate customers. Phil found that some customers got behind in their payments and were ashamed to come in to the business. Instead, they paid cash at a competitor's.
- Credit creates business. If everyone demanded cash transactions, many sales would disappear. At the least, business would be slowed. Credit builds volume, Figure 15–1.

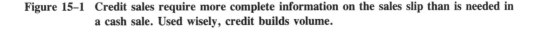

	R. J. Marshall 18 E. Martin Drive					
	Retail Sporting Goods Inc. Highland, NY 12528					
SOLD TO_____				INVOICE NO _____		
_____				DATE _____ 19____		
_____						
TERMS _____			SHIPPING VIA _____			
QUANTITY	DESCRIPTION		UNIT PRICE		TOTAL AMOUNT	

Figure 15–1 Credit sales require more complete information on the sales slip than is needed in a cash sale. Used wisely, credit builds volume.

MANAGING CREDIT

What are the Advantages of Providing Credit at the Point of Sale?

1. One-stop shopping is possible, thus saving customers' time.
2. Ease of buying or convenience will increase sales.
3. Credit information and records can be a help in planning special purchases and promotions.
4. Credit may bond customers to the firm because of the credit support.
5. Credit customers may be more quality and service oriented than price motivated.
6. Goods can be sent on approval.
7. Credit is a useful sales tool.
8. Because of credit extension you may know the customers better.
9. Credit builds goodwill when kept current.
10. Credit provides a way to earn a return on excess funds when carrying charges are added to sales.

What are the Disadvantages of Providing Point-of-Sale Credit?

1. Costs of doing business are increased—postage, bookkeeping, interest, and time.
2. There is a need for more capital, and capital may be limited anyway.
3. Credit usually risks some losses. If credit is used effectively, there will be some bad debt losses.
4. Some credit customers have the intent to defraud.
5. Time must be spent in checking the credit record of customers.
6. Economic cycles may result in unusual credit losses.
7. Some credit customers stretch their payment ability by not allowing for unplanned events (livestock disease, drought, or hail, for example).
8. Credit customers are more likely to return items and request ''on approval'' privileges.
9. Credit offered costs the average business from 3 to 5 percent of the sale.

What Are the Types of Credit?

Open Charge Accounts.   This type of credit is handled by the business and is characterized as being due upon billing, usually on a 30-day cycle. The billing may have a percentage discount if paid in 10 days—net in 30 days with a carrying charge added after 30 days. Figure 15–2 shows key elements for credit management.

Easy Payments or Installment Credit.   These accounts are set up to schedule payments for durable goods such as appliances, furniture, or other items that would be salable upon repossession. There is more protection for the retailers than with the charge account. Usually a sales contract and down payment are required, and in some cases the credit instrument is sold to a third party.

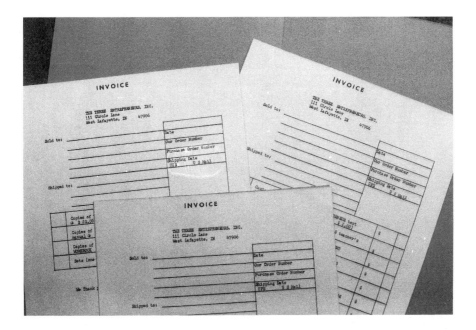

Figure 15–2   Invoices and statements are at the core of a credit management system.

Store Revolving Charge Accounts.   Many large department store chains provide their own credit cards for purchases made in any of their stores nationwide. With this type of account, the customer is billed monthly for purchases made during the month. The customer can pay all of the balance with no finance charge, or they have the option of making installment payments with a finance charge added on amounts over 30 days old. Many of these accounts have a credit limit and the amount of difference between the balance due and the credit limit is available for additional purchases. After successful experiences with the issuing company, it is possible to have the credit limit increased by negotiation.

Bank Cards.   "Plastic money" or bank cards are accepted by many businesses. For a fee, the bank issues the card and bills the customer at regular monthly intervals. Prompt payment costs the buyer nothing but when only part of the balance is paid, the customer pays a 1–1½% per month finance charge on the average daily balance. A business benefits in two ways. First, the credit is advanced by someone else and second, someone else keeps the credit records. This service costs the business a small percentage of each credit card transaction.

Business Use of Credit Cards

Many businesses use credit cards to pay for services. For example, gasoline credit cards permit business employees to charge travel expenses while on company business. Monthly statements provide a detailed record of expenditures without the hassle of advanced payments and return accounting through expense accounts. Cash business, however, avoids many credit problems. The Copymat, Figure 15–3, carefully controls credit.

Figure 15-3 The Copymat does little credit business. (Courtesy Don Steele)

Consumer Credit Regulations

The truth-in-lending law applies to bank card accounts. It requires disclosure—a statement of the conditions under which a finance charge may be made, how the balance is to be charged, how the finance charge is determined, the minimum payment required, the interest rate charged, and its corresponding annual percentage rate. In addition, the disclosure statement must tell how any lien may apply on the customer's property such as the right to repossess an appliance or other purchase.

How Credit Sales Affect Businesses

As noted earlier, credit provides the opportunity for many consumers to purchase needs and wants. If people were required to pay cash for TV sets, many would have a great deal of difficulty accumulating the $550–700 price of a 25-inch console. When credit is extended, the customer finds it possible to pay $50–60 per month and so is willing to buy the set.

Financially it may appear as follows:

Purchase price	$625.00
Less the down payment of 10%	62.50
Balance due	552.50

But this store's finance charge is 10% of the purchase price, leaving the unpaid balance of $625 to be paid in 12 equal payments.

Month	The Unpaid Balance at the Start of the Month
1	625.00
2	572.92
3	520.83
4	468.75
5	416.67
6	364.58
7	312.50
8	260.42
9	208.33
10	156.25
11	104.17
12	52.08

Many customers do not realize the high cost of interest in such purchases. The actual cost is the sum of the first balance plus the last balance divided by two, resulting in the average amount financed over the period of the payments.

$$\frac{625 + 52.08}{2} = 338.54$$

When the finance charge is divided by this amount:

$$\frac{62.50}{338.54} = 18.46 \text{ percent actual interest paid}$$

If the retailer has idle money, this is a way to increase returns on funds. If, on the other hand, money is limited, the dealer must borrow funds to support the credit extended or sell the contract to a third party at a discounted price. The third party, then, collects the major portion or all of the finance charge.

Credit Applications

Credit applications are used by most people who extend credit as a way to evaluate the degree of risk involved in extending that credit. The amount of risk is balanced against the volume of business created by taking the risk. Each company has its own formula to determine the level of credit risks to be extended. Bad debt losses or write-offs are a part of this formula in most companies. Credit investigations are conducted by most credit departments. In major cities, credit bureaus are available to provide such credit histories to members of the bureau.

Collections

Survival of businesses who extend credit may depend upon the ability of the credit department to make collections of *accounts receivable* in time to maintain a positive cash flow.

Credit-granting businesses need to continually monitor credit and become aware of slow accounts. Monitoring must be updated on a monthly basis and policies formulated to manage credit.

Using the Telephone in Collections.   Phil used the telephone in collecting his oldest bill by making a definite appointment to talk to the customer. He was careful, however, not to discuss credit on what was a party line. The telephone can permit immediate feedback concerning the credit items. Questions and answers are possible in determining the possible problems.

Collection Agencies.   When your collection efforts fail, you can hire professional collection agencies. You pay them a percentage (usually large) of the amount collected for their expertise.

Establishing a Credit Policy for Your Business

There are a number of important questions and concerns that need to be considered as you decide whether or not to use credit as a sales tool. Among these items are the following:

- Your cash flow position.
- Your sales projections with or without offering the credit.
- The ability to recover the product in case of nonpayment.
- The reputation of the potential customer.
- Your capital availability.
- The ability to resell the credit instrument.
- The possibility of an agreement with a credit card company or companies.
- The best terms to establish as you issue credit.
- The margin of profit on the goods offered for sales.

Your financial position should help you make a decision on credit policy. If you are borrowing operating capital, as Phil was, credit doesn't appear to be a wise choice. Cash flow problems can sink a business in the short run when the business is sound in the long run. Accounts receivable won't meet the payroll or pay the *accounts payable* in a timely manner.

Credit sales become accounts receivable and each month they cost the business money. The load of corn sold by Phil to Joe Hearty's son ended up costing Exeter Feed and Grain over four times what the net income would have been had it been a cash sale. In the low margin feed and grain business, accounts not paid in 5 to 6 months result in net losses and hurt the business more than they help.

Your margin is an important factor in the credit decision. With a large margin, credit is a better sales tool than it is in a low margin business. In order to be profitable, the finance and carrying charges must exceed the cost of interest and the expense of collection, accounting, billing, and bad debt losses.

Type of merchandise is another important consideration. Durable goods, which have a high value even when used, are more logical items to allow as credit sales when compared to consumable or nonrecoverable items. You probably would not want to extend credit for items that are of no value after first use. Consumables such as feed and grain may be sold on credit since the livestock can be a *chattel* for their sale.

A credit policy should stipulate the conditions for the sale, method of payment, length of the credit contract, finance or interest charges, and the person making the decision on whether to extend credit. In addition, credit contracts should specify the method of collection and costs of collection when the agreement is broken. The right or conditions under which the contract can be sold to a third party may also need to be stated.

Credit experience may result in a customer receiving special considerations or policy exemptions, or conversely being denied credit.

Credit card companies charge a fee ranging from 2 to 5 percent on each credit card transaction as their price of handling the credit. Under this agreement, businesses forego some portion of their income on credit sales in order to transfer the accounting responsibilities and to improve cash flow.

As noted earlier in the chapter, credit may be a profitable use for excess cash. When the business is able to invest its extra capital, it can better its income through increased sales and interest collected on accounts receivable.

In the start-up phase of the business, the effect of credit on gross sales will be difficult to estimate. As the business operates, the number of persons who inquire about credit may give you a way of estimating the potential for increased sales through credit extension.

Using Credit as an Entrepreneur

As an entrepreneur, the ability to obtain credit may be a valuable resource. Quantity purchases, purchases where repayment depends on the first sales, and other similar situations magnify the ability of the small business to grow and prosper in a competitive climate. As a small business manager who is just starting, some credit guidelines are needed. For example, five months after the crisis detailed in this chapter, Phil closed out the year's business with a large percentage of accounts receivable being paid up by farmers who reported income taxes on the cash basis. He quickly paid off the supplying mill and requested a review for a higher credit line in order to be prepared for the next summer/fall's credit crunch.

SUMMARY

Credit is a sales tool that can help a business grow but can also cause it to fail. Credit management is an important activity for credit-granting businesses. There are credit bureaus and credit investigators that can help businesses control credit risks.

A no-credit policy may well be the best way to start in a business. Currently, there are many businesses that operate only on the cash basis, and many others who arrange to accept bank cards, thereby putting credit decisions on the third party, the bank. The acceptance of checks will be covered in another chapter.

SELF-EVALUATION

Credit Policy Quiz

1. List six ways credit can help a business.
2. List six ways credit can hurt a business.

3. How do you determine the actual annual percentage rate of interest on a 12-month loan?
4. What elements do you think are most important in a credit policy?

STUDENT ACTIVITIES

1. What is the APR when the finance charge on a $750 TV set is $75.00, the payments are to be $70.00 per month, and no down payment was asked?
2. Ask stores that grant credit how they make credit decisions.
3. Contact the local credit bureau to find out how it operates.
4. Obtain a credit application, complete it, and then have a class member act as a credit manager and evaluate it.
5. Ask your classmates how many favor credit and how many oppose it.
6. With your class debate the issue, "Resolved, credit helps a business more than it harms the business."
7. Develop and solve several APR problems to better your problem-solving skills.

CHAPTER 16

How Do I Manage the Legal Aspects of a Business Venture?

OBJECTIVE

To identify some of the legal items an entrepreneur needs to be concerned about when starting a business enterprise.

COMPETENCIES TO BE DEVELOPED

After completing this chapter you will be able to:

1. Identify common legal requirements for a beginning business.
2. Recognize the need for a business attorney.
3. Recognize the need for governmental services.

TERMS TO KNOW

Building permit	Copyright
Easement	Highway access
Lease	Legal advice
Licenses	Local code enforcement
Options	Ordinances
Occupational Safety and Health Act (OSHA)	Patent
Patrons	Property taxes
Sales tax	School levies
Tax withholding	Unemployment insurance
Vendor	Work permit
Worker's compensation	Zoning laws

INTRODUCTION

Many laws have been passed to safeguard the interests of the public and other groups. Some have worked well to protect the general population and others have only served to protect some groups from any competition.

CASE STUDY

The Case of the Building Permit

Cedric started the foundation for his new barber shop in a small rural community and was informed he needed to have a *building permit.* He drove to the county courthouse, paid the five dollar fee, and went back to lay the floor joists and subfloor. A day or two later the county building inspector stopped by to check the progress. The good news—he had not backfilled the foundation footings and they were usable. The bad news—he had failed to put a termite barrier at the top of the foundation wall. Fortunately, at that stage he was able to loosen the tie down anchor and lift the floor to install the needed metal guards. At that time he obtained legal help with the local building code to avoid further costly errors.

What We Learn From the Case

- One must know the local, state, and federal rules and regulations governing business activities.
- In areas where you don't know these rules, *legal advice* is advisable.
- Most business enterprises have many rules and regulations that govern activities.
- It is best to be informed before the fact than to learn by penalty.

THE LAW AND YOUR BUSINESS

Among the legal requirements one must consider in starting a new business are many that everyone will recognize: federal, state, and local income taxes, real estate and *property taxes,* city and state licensing fees, *tax withholding* rules, *sales taxes, worker's compensation,* and social security taxes. Other legal items are not as familiar. These include *options,* zoning rules, building permits, OSHA regulations, *easements, work permits, patents, copyrights, leases,* and local *ordinances* pertaining to such things as signs and parking spaces.

Tax Withholding

Businesses are required by law to withhold income taxes from employees' wages. These withholdings may include city or county taxes in addition to state and federal income taxes, and social security taxes. Employers need to obtain identification numbers and keep the amounts withheld in a separate account until the required reporting period. The frequency of deposit or payment of these funds is determined to a certain extent by their amounts. Any payments made after the deadline are subject to a penalty.

Sales Taxes

Many business ventures are required to collect sales taxes on their sales of goods and services. How these laws apply vary from state to state just as income tax laws do. Again, there are reporting deadlines and penalties for late filing. The Internal Revenue Service requires a tax identification number for all businesses. Obtaining this number should be one of your first actions as an entrepreneur.

Real Estate and Property Taxes

Taxes on a portion of the value of real property and business inventory are common in many states and localities. These frequently fund schools and local governments. Some states allow school *patrons* to vote on the proposed *school levies.*

Leases

Leases are familiar to most people as rental agreements on real property. Leasing has grown in popularity and is currently extended to use of office equipment, computers, and delivery trucks. Leasing requires less capital than ownership and is used more frequently as a means of extending a business's potential capital base.

Options

There are several types of options that are important to businesses. An option is a fee paid for the privilege of selling or buying a commodity or property at a fixed price until a certain future date. For example, a business is considering building a new plant. In order to protect a desired property location as the site for expansion, the company will take an option on the location at a guaranteed price.

Worker's Compensation

Worker's compensation is an insurance policy provided by a private carrier and paid for by the employer to protect his employees from certain job-related health problems and accidents.

Businesses with a certain number of employees are required to contribute to *unemployment insurance.* It is unlikely that these rules will apply to you unless your enterprise grows very rapidly.

OSHA

OSHA, the Occupational Safety and Health Act, sets safety standards for businesses with a minimum number of employees. The purpose of the act, controlled by the Department of Labor, is to maintain a safe workplace for all employees.

Copyrights and Patents

Copyrights and patents are two ways of protecting an original work. Patents apply to new inventions while copyrights apply to written materials, music, computer software, and other similar items. Copyrights and patents can be obtained by individuals or corporations, but the services of a lawyer familiar with these laws should be sought.

Licenses

Licenses are required by many city, county, state, and federal authorities. Licenses are viewed as a means of controlling or regulating the services given by professionals or the services of common carriers such as taxi cabs and buses.

Ordinances

Many communities regulate certain types of businesses by means of local ordinances. For example, most cities will not allow businesses selling alcoholic beverages to be located near a school. Another common ordinance pertains to burning leaves or refuse within city limits.

Zoning Laws

These are special ordinances that pertain to zones or zoning. They regulate the location of any type of business. One zoned area may be approved only for single family housing, while another may permit multiple housing units to a certain density. Other zones limit local business, light or heavy industry, and, important to all of us, agricultural use. Most of these *zoning laws* can be appealed to a zoning board or commission. An entrepreneur will need to be aware of these laws when he starts a new enterprise. Figure 16–1 shows a request for variance.

Building Permits

Many communities have building codes. To assure safe construction and preserve aesthetics, among other reasons, building permits are required in order to construct or alter existing buildings, Figure 16–2. A permit is issued and each phase of construction is inspected to ensure conformity with local codes.

Easements

Easements are documents that provide right of way for utilities and in some cases access a business driveway across someone else's property. An entrepreneur would need to know about easements when building a new facility.

Work Permits

Work permits are needed for people under the age of 18 if they are to be employed in a business enterprise on a regular basis. Students obtain these, in most cases, from the local school administration.

Figure 16–1   A notice of request for rezoning is posted at the site.

Figure 16–2   Legal permits are required for building or improving a business site. (Courtesy Creekside Animal Hospital)

Signs

Many cities, counties, and states have laws relating to the size and location of a business sign. For example, a young man purchased a restaurant and then was sold a sign to replace the one on the roof of the building. Later he discovered that the old sign was nonconforming and therefore could not be replaced. Small claims court received that problem because the sign salesman was guilty of misrepresentation. Figure 16-3 shows a legal sign in one community.

Figure 16-3 Signs are regulated as to size and location in many communities. (Courtesy N. J. Plantenga)

Parking

Some communities regulate the number of offstreet parking spaces that must be provided. Most businesses realize the need for adequate parking space as a business attraction.

Highway Access

Highway access is important to most businesses located near major highways. With many limited access highways constructed in recent years, access roads have become important in the sale of real property.

Health Department and Other Regulations

You should always become familiar with local rules and regulations that pertain to your business. For example, a business built outside city limits, without sewer connections, will need an approved septic system and the county health department will have to approve the site.

Where does one get such information? In most units of government, there is an official charged with overseeing *local code enforcement.* This person can supply information and guidance on meeting code requirements.

It is important to obtain written documentation of what is required. These written requirements (or specifications) should be included in any contract for work. In one case, completed work had to be torn out and redone to meet code. This is a costly process!

The example cited in the paragraph on signs points out the need to get firsthand information. Sales people or *vendors* are not the best source of information. Their interests are not necessarily yours.

THE NEED FOR A COMPANY LAWYER

The number of different items a business venture must keep track of points out the necessity of retaining legal counsel. An attorney will be able to handle many items for the business, from options and leases to taxes and advice on organizational structure.

SUMMARY

Business activities are governed by a maze of rules, regulations, and laws that are supposedly designed to protect the public. The cost of keeping up with these rules is often passed along to the consumer and taxpayers in the form of higher prices. In many cases the costs are less than if no regulation occurred.

SELF-EVALUATION

Legal Quiz

True-False: On a separate sheet, write T for true statements and F for false statements.

1. All laws benefit the general public.
2. Ignorance of a law is no excuse.
3. Businesses are not required by law to withhold income taxes.
4. Businesses have relatively few restrictions on their operations.

5. Businesses are paid for collecting sales taxes.
6. All sales by a retail business are subject to sales tax.
7. Many states collect property taxes from businesses.
8. Leasing equipment may aid a business by providing operating capital availability.
9. Copyrights and patents are designed to protect original ideas.
10. Work permits are designed to protect children.
11. A business may have restrictions on the location and size of outdoor signs.
12. Written records of any legal action are worthwhile records.
13. A company lawyer can be a valuable asset to an entrepreneur.
14. Parking space is a necessity for most businesses.
15. Most rules and regulations cost consumers while protecting their interests.

STUDENT ACTIVITIES

1. Obtain a copy of local ordinances and look for those that affect business.
2. Find out what local regulations would apply to signs you might put up to advertise your business.
3. Obtain a copy of local building permits or other permits you would need to start your type of business.
4. Contact a business to find out what operating permits they have obtained.

CHAPTER 17

How Can I Manage Tax Responsibilities for the Business Venture?

OBJECTIVE

To identify how taxes affect the entrepreneur.

COMPETENCIES TO BE DEVELOPED

After completing this chapter you will be able to:
1. Identify the kinds of taxes that apply to an entrepreneur.
2. Identify correctly the following terms:
   - user fees
   - property taxes
   - income taxes (local, state, federal)
   - sales taxes
   - excise taxes
   - social security taxes

TERMS TO KNOW

Accounting system
License fee
Property tax
Remitting
User fee

Excise tax
Procrastination
Recourse
Taxation

INTRODUCTION

Society has chosen taxes as a way to fund general community services. Individuals and businesses are taxed to raise these funds. In this chapter we look at the nature of taxes, their common forms and their effect on businesses.

CASE STUDY

Phil Pays a Penalty or The Case of the Improper Postmark

Phil's daily business summary had a column for collected sales tax. Each monthly summary totaled these daily totals and once each quarter a report was filed with the state internal revenue department *remitting* the taxes collected with the required report. In early July, Exeter Feed and Grain had been very busy and Phil delayed completing the report. He delivered the report to the post office just before closing on the day it had to be mailed. The report was postmarked the next day and several days later Phil received a bill for the late penalty.

Naturally, Phil was irritated because the report was mailed on the required day. However, he paid the penalty as there was no *recourse*. He resolved that in the future he would do the reports earlier in the month and be sure to have the letter postmarked before leaving the post office.

What the Case Tells Us

- Taxes and tax reports must be filed on time.
- *Procrastination* can be costly.
- Timeliness in reporting is good business.
- People are unaware of urgency unless concern is shared.
- Reports done promptly take the same amount of time and prevent later stress.

LOOKING AT TAXES

Society, over the years, has reached the conclusion that certain services should be performed in the public good. Taxation provides a method of distributing the costs of these services among the public. Some of these services are local or city in nature, others are township, county, or statewide in scope, while others are national in their effect. The types of services we are most familiar with are police and fire protection, the operation of our schools, sanitation service, government operation, parks and recreation programs, and highway maintenance.

There are many different types of taxes used to raise the needed funds. The more common taxes currently in use are *user fees, excise taxes,* federal, state, and local income taxes, and permit fees. Each of these taxes has a tax base and a service it supports to a certain degree.

Your school, for example, is supported locally by *property taxes*. Statewide support may have come from multiple sources. Local government services are likely to be funded mostly from the property taxes levied on real property, personal property, and business inventory.

Examples of user fees include state and national park entrance fees, and support of the highway system through gasoline taxes. Some city governments are dealing with the budget squeeze by charging for sanitation services and trash pickup, services that once were supported by taxes.

Further examples of tax-supported services include licensing of beauty operators, taxis, and so on. Funds raised through license fees are used to protect the public by maintaining standards.

ARE TAXES FAIR?

Loud debate rages over the fairness of any given tax. The recent tax overhaul in Congress is a good example. What is considered fair grows out of our heritage and since our heritage as individuals is so diverse, we have many opinions.

In a family, the parents determine what is fair. In this country the political process determines policies that dictate our equity policy. Income taxes were first assessed on the theory that those most able should pay more. Property taxes were assessed on the basis of what was owned. Laws have been modified over the years to adjust what some groups have thought to be favoritism for other groups.

Economics is the efficient allocation of resources. Taxation in theory is derived from this concept. In a complex society, as in an imperfect world, the two are not always smoothly meshed.

Whether or not taxes are fair depends on your viewpoint. It is clear that what one person considers fair, another will not. At best, taxes and tax rates are a compromise. Tobacco users don't all favor the tax on tobacco while nonusers think the taxes should be higher.

What Kinds of Taxes Do We Need to Consider in Operating a Business Venture?

Every business person needs to be aware of the taxes, licenses, permits, and so on required in that particular type of activity. Some of the common taxes are taxes on real property, business inventory, sales taxes, social security taxes, food and beverage taxes, hotel taxes, excise taxes, worker's compensation, and all the licenses and permits required in business today.

Your attorney, the city clerk, the county treasurer, the State Bureau of Internal Revenue, and the Secretary of State's offices are just a few sources of information on taxation and licensing. An inquiry to any of these sources will likely lead you to additional sources.

Ignorance of the tax laws in your state will not excuse the nonpayment, nonfiling lapses of a new business. Most failures to report or obtain the correct permits will carry penalties, and many of these are much more expensive than obtaining necessary permits and paying the taxes on a timely basis. In fact many new entrepreneurs have found that once they make a mistake that is caught, they seem to be audited on a regular basis.

Businesses are required to withhold income taxes from employees' wages and deposit these in special accounts at specified intervals. These intervals depend upon the amount of taxes collected in a given period of time. Social security taxes are collected in the same manner.

Property taxes are assessed on a given date each year and are based on the valuation, which is determined in a specific manner. Most states determine the assessed valuation at regular intervals in an effort to determine a fair basis for taxation. Business inventories are assessed on a given date (such as March 1). Where this system is practiced, businesses try to reduce their inventories before that date by passing along a large portion of the tax saving as a buying incentive.

**Figure 17-1 Dave and Tammy Buck give their records extra attention at tax time.
(Courtesy David Buck)**

Excise taxes are collected on various products. The costs and returns of these taxes are hotly debated. Prime examples of these taxes are the ones levied on tobacco products and other goods considered of little social value. People question the use of tax revenues for anti-tobacco education programs, when tax revenues also subsidize tobacco growth. In fairness, however, excise taxes raise billions more than the tobacco subsidies cost the government.

Tax laws are complex and many small businesses choose to employ a tax specialist to complete their tax returns. Taxes should be considered when writing the business plan. Taxes also should be considered when accounting practices and records are established for a new business venture. A little time focused each day may save hours at tax reporting time. The *accounting system* can help simplify these reports.

SUMMARY

Taxes are an important concern for the entrepreneur beginning a new business venture. The entrepreneur must find out how the tax laws affect the intended enterprise. Responsibility for knowing the tax laws rests on each business owner or manager.

Taxes are assessed in order to pay for governmental services to our society. Services such as police and fire protection are more effectively served by public rather than private funding. Taxes are collected on the basis of some common factor such as property values (real estate and property taxes) or purchases made (sales taxes). Careful attention to taxes is important for the entrepreneur as the Bucks recognize, Figure 17–1.

SELF-EVALUATION

Tax Quiz

Multiple choice: On a separate sheet, write the letter representing the best answer to the stem question.

1. Taxes are a recognition that
   a. private enterprise cannot do everything.
   b. society needs to do certain things collectively.
   c. users should pay for their services.
   d. permit fees are useful.

2. User fees are taxes or fees assessed to
   a. the general public.
   b. entrepreneurs.
   c. provide sanitation services.
   d. users of a service.

3. Property taxes are based on the theory that
   a. those most able should pay.
   b. the owners of property should pay.
   c. users of a service should pay.
   d. favoritism should be shown to special groups.

4. Income taxes were started on the theory that
   a. those most able should pay.
   b. there should be equal support of services.
   c. users of a service pay.
   d. everyone must support the government.

5. Sales taxes are taxes assessed on
    a. park entrance fees.
    b. salaries and wages.
    c. most purchases.
    d. a person's income.

6. Most public schools are supported by
    a. excise taxes.
    b. income taxes.
    c. user taxes.
    d. property taxes.

7. User fees are illustrated best by
    a. excise taxes.
    b. income taxes.
    c. gasoline taxes.
    d. property taxes.

8. Unemployment taxes are designed to
    a. protect employees.
    b. provide for injury or accidents.
    c. provide for survivors of industrial accidents.
    d. fund OSHA activities.

9. Entrepreneurs are subject to a number of tax reports. The reports should be
    a. turned over to an accountant.
    b. made on a timely basis.
    c. paid for making these reports.
    d. made under protest.

10. Taxes are assessed in a number of different ways in order to
    a. spread out the costs of government.
    b. pay for the mayor's salary.
    c. pay for most government services.
    d. pay for the state government.

STUDENT ACTIVITIES

1. Interview a tax preparer to find out how business taxes are different from individual returns.
2. Ask your local county treasurer what taxes are paid into the county, how they are assessed, and how the funds are used.
3. Interview an IRS person to find out how tax audits are completed.

CHAPTER 18

Where Should My Business Be Located?

OBJECTIVE

To explore the importance of location to a business.

COMPETENCIES TO BE DEVELOPED

After completing this chapter you will be able to:
1. Correctly identify six major concerns in selecting a business location.
2. List the reasons for each concern.
3. Make an informal traffic survey as it relates to business activity.

TERMS TO KNOW

Chain	Clientele
Competition	Confront
Congested	Inside information
Isolated	Spin-off
Trading area	Traffic

INTRODUCTION

Location has long been recognized as one of the important factors in business success. In this chapter we look at the factors that influence the quality of a business location, from choosing the city to considering local ordinances and personal preference.

CASE STUDIES

The Case of the Missed Location

A major *chain* of stores had an executive that worked closely with the state highway commission. A new bypass was planned for one of the state's major cities, and as soon as the plans were drawn, the chain company bought land and erected a store. However, when the planned highway improvement was exposed to public comment, it was intensely opposed. The highway bypass plan had to be dropped. The store wasn't successful in the otherwise *isolated* location and closed after two years of losses. What looked like an outstanding location with the planned highway interchange turned out to be a small country road without adequate *traffic* to sustain the store.

At about the same time, the chain started another store at another highway into the city. The location turned out to be part of a growing business area. In a few years a development company opened a shopping mall a few blocks away. The company closed their store and moved into the mall to benefit further from the additional traffic generated by the other retail outlets.

The Case of the Better Location

Exeter Feed and Grain was a *spin-off* of the Exeter Motors implement dealership, which was started originally by a major auto dealership located on the town's square. The new business was located on the main road into town coming from the bypass.

The major *competition* in the grain business was an elevator located on a side street. It was soon evident that the main street location was much busier than the one on the side street.

What the Case Studies Tell Us

- All other things being equal, the higher traffic site is probably the better location.
- *Inside information* does not always pay off.
- Location is influenced by a multitude of interrelated factors.
- Location can make the difference between profit and loss for a business.
- Visibility is an important factor in location.

LOOKING AT LOCATIONS

What makes a good location for a business is a topic discussed by business owners around the world. Some of the key issues may cause the business to prosper or to fail. Among the considerations in determining a business location are the following:

- Choosing the city or town
- Choosing the area within the city or town
- Evaluating traffic patterns
- Choosing a building or facilities to fit the nature of the business

- Evaluating the competition
- Considering growth and expansion
- Site costs compared to projected returns
- Local ordinances, laws, and regulations affecting business location
- Personal preferences
- City zoning ordinances

Choosing the City or Town

A major factor to consider in selecting the city or town for your business enterprise is its general economic conditions and its *trading area*. The author once lived in a rural town that had a very large trading area. People actually drove through the county seat town in order to trade in the slightly smaller but more active town. It would be wise to select a growing, dynamic area as compared to a declining town as a business location.

The presence or absence of a strong merchants' association can be of value in this decision. A group of community-minded promotional merchants organized in some manner can help promote business for everyone. They can also be excellent sources of information useful in decision making.

Choosing the Area Within the City or Town

The decision of where to locate in the city or town is probably much more important than the choice of city or town. The general area would be more important to a wholesale business, while the more numerous retail businesses would find the specific area of town most important. Figures 18–1 and 18–2 illustrate good and poor locations.

Exeter Feed and Grain found an excellent location on the main traffic artery that provided visibility and highway access. The founder of Exeter Motors had the foresight to provide for potential growth by obtaining a 10-acre site at the time the bypass was constructed.

In choosing a location in the city, it is also important to study the normal traffic patterns of your potential customers. To illustrate this, a business catering to elementary school children's needs would benefit from a location en route to the elementary school. An auto-related business would thrive on "automobile row" near the auto dealerships. Needless to say, seedy, rundown areas of the community could prove to be a handicap that would be difficult to overcome.

Evaluating the Traffic Patterns

The ability of customers to enter the business premises will have an effect on the business. Traffic may be so *congested* in certain parts of the city that people have difficulty crossing the street to enter. In such cases, access from two streets, as on a corner location, may improve business potential. Consider also the other extreme—a remote location with such sparse traffic that very little potential business exists.

Closely related to traffic is the parking problem. Downtown areas have declined rapidly in recent years because of restricted parking areas (parking meters and paid lots). Even with validated parking, people will go to mall areas to find adequate free parking and enclosed

Figure 18–1   This location was evidently a poor one for the former occupant.

Figure 18–2   This is a good location. The business is doubling its space. (Courtesy Creekside Hospital)

walkways. Adequate parking is essential for good business traffic. A grocery store with inadequate parking will be bypassed in favor of the competition with larger lots.

Choosing a Building or Facilities for the Business

The appearance of a business has an effect on the type of *clientele* as well as the number of customers who patronize it. Recognizing the limited resources of most beginning entrepreneurs, our advice is to make the business as attractive as possible while staying within the business plan's budget.

The building has to meet or allow for meeting the basic needs of the business. A business selling bulky items would need inside and outside doors large enough to permit handling of the merchandise. A business selling heavy appliances would need a loading dock. An engine rebuilding business would need an overhead hoist and track, so the building framing would have to support this.

The nature of the business also affects the location and choice of facility. For example, a sand and gravel business would need to be near the source of these materials for economy of handling. A source of water may also be required in order to clean the silt and organic matter from the sand.

Evaluating the Competition

An examination of the competition for a proposed new business needs to be made before deciding on the location. In larger centers, it might be better to move to another major traffic area than to confront existing businesses in the same area. The different location would enable the new business to attract customers that otherwise might not go to the competition's location.

Evaluation of competition would also be involved with other business decisions that are discussed in other chapters.

Location Costs as Compared to Potential Returns

An ideal location will attract many different businesses and thus may be in a high rent area. Such costs are important in projecting the cash flow of the business plan. Careful evaluation of location costs versus anticipated sales is essential in deciding the optimum location. On the other hand, a low rent location may be so unfavorable that business may not even warrant the lower payment.

A few years ago an Ohio department store manager left his job and started a hobby supply and plant store two blocks from the main street in an area with few shops. Several creative entrepreneurs built this into an excellent business area. The landlord, seeing how they prospered, raised the rent 50%. Despite loud complaining most of the businesses stayed because they realized it was more profitable than moving and again rebuilding store traffic.

Other Businesses in the Area

The number and types of businesses in the area must be considered. The example of the auto-related business above illustrates one facet of this concern. A second consideration, along with related and noncompeting business, is related and competing business.

Noncompeting related ventures will usually build business but competing ventures will usually divide business. An example of this occurred in an Indiana city a few years ago. A new fast-food chain moved into town and took the best available location, which was cramped for parking space and for future expansion. As the business grew, the building was enlarged and soon the parking lot was full during the lunch hour. With both a front and rear entrance to the site, one could observe people driving on to the lot, seeing no parking available, and driving across the street to another chain with a larger parking area. The neighboring business profited considerably on this second choice basis.

Meanwhile, during its growth and establishment time the business next door foundered, and a third chain snapped up the location, which was also too small for an adequate franchise. The first business missed the chance to provide for expansion and growth.

Considering Growth and Expansion

The potential growth and expansion of the business should be considered in locating a business. Although it will not matter in the short run, it will matter greatly over the long run. When a business is relocated, customers will be confused and dislocated for a period of time and this will result in lost business.

Site Costs as Compared to Projected Returns

Purchase price or rental costs can give you part of the information needed in attempting a financial analysis of a proposed site. The other half of the equation is harder to determine. When the projected net revenues at one site meet or exceed those of a second, then other factors should be examined more closely. Market study needs attention at this stage in the decision process.

Local Ordinances, Laws, and Regulations
as They Affect a Business Location

Another factor to consider in choosing a location are local ordinances, laws, or regulations. Many cities have ordinances designed to prohibit certain business in areas considered inappropriate by the community. For example, in many cities a business with a liquor license is prohibited within a certain distance of a school. Also, no businesses may be located in areas zoned residential. Many additional restrictions apply to businesses at various locations.

Personal Preferences

Personal likes and dislikes should be considered in decision making, but should be tempered by good sense. You may have a family reason for choosing a particular location. For example, a man's grandfather gave him a property, which reduced his initial cash outlay in starting a business. It was a big help in the short run but a hindrance over time due to its inadequate location. In the long run, the business survived without really prospering until a relocation and expansion took place. Most professionals advise against letting personal preference influence a location.

City Zoning Ordinances

Finally, consider local zoning ordinances. They may be quite specific about which types of businesses can locate in certain areas. If restrictive ordinances apply, one can appeal and sometimes win approval.

Another area of concern in many parts of the country are restrictions on business signs. A young entrepreneur who did not know about restrictions bought a replacement sign for his roof only to find he could not install it—a costly mistake.

SELF-EVALUATION

Location Quiz

True-False: On a separate sheet, write T for true answers and F for false answers.

1. Inside information is always reliable.
2. Inside information may give a business the jump on a competitor.
3. A higher traffic site is usually a better location.
4. Parking is always more important than location.
5. A corner location with parking is superior to a middle of the block location.
6. Personal preference may outweigh all other location factors.
7. Shopping patterns may be more important than the size of the city selected.
8. Location in an area of related business will help build store traffic.
9. A desirable location for your business may be desirable for other businesses as well.
10. Future expansion is a valid location consideration.
11. Expansion possibilities are important in the short run.
12. Zoning is not important in selecting a location since it can be changed.
13. The possibility that a later move could cause business interruption should govern site location in starting a business.
14. One should check out local ordinances such as street setback, parking spaces, and sign restrictions in determining a business location.
15. Location is always more important than financing during the business start-up.

STUDENT ACTIVITIES

1. What stores in your earlier survey are well located? Poorly located? Why do you classify each in this manner?
2. Assume you have just been granted a franchise of one of the following types of business. You are to open July 1. Where would you locate in your community? Why?
   a. fast-food
   b. hardware
   c. income tax service
   d. clothing store
   e. lumber yard
   f. trucking company
   g. gas station

3. Check the traffic flow at any vacant corner in your city or town, then check the traffic flow at the corner of a successful business and compare them.

4. Compare the traffic flow at the sites of two competing businesses.

5. Assume you have decided to go into a gas station business and are offered two different locations. The first is at the intersection of the two main highways at a rent of $1,000 per month; the second is located eight blocks away from one of the highways and is offered at $500 per month. Which would you choose and why?

CHAPTER 19

What Facilities Do I Need For the Business?

OBJECTIVE

To consider the facilities required by an entrepreneur.

COMPETENCIES TO BE DEVELOPED

After completing this chapter you will be able to:
1. Identify utilities needed to fit a business location and type.
2. Identify the problems a business move will create.
3. List the advantages of renting or owning the business facilities.
4. Evaluate a location's suitability for a planned business.

TERMS TO KNOW

Adaptations	Facilities
Setback	Structure

INTRODUCTION

What facilities do I need for the business? The answers to that question are as varied as the owners who start businesses. Figure 19–1 depicts an available business location. This chapter will examine a way of looking at what *facilities* may be required. At the lower end of the range, many profitable businesses have consisted of a mailing address and a telephone line. At the other extreme, a copper mine covers thousands of acres and utilizes massive equipment. Entrepreneurs must plan well for facilities.

Figure 19–1   This business building is available. Before renting, the prospective tenant should check the facility for suitability.

CASE STUDIES

The Case of Needed Expansion

When Phil became a partner in Exeter Feed and Grain, the proposed site included a 20' x 20' building, office, a 15,000 bushel elevator, and a 100,000 pound scale located on the Exeter Motors acreage. In addition, an 1,800 square foot warehouse attached to the elevator provided adequate feed storage. Since the sanitation stock and sale supplies would not fit in the 20' x 20' building, the building was doubled in size to 20' x 40' with suitable display areas and stock shelves added. Major-sized feeders were displayed outside the building and supplies, except for display items, were stored in the feed warehouse.

The Case of a Forced Move

An economy clothing store had occupied a low-rent building for several years and had become established as a place to obtain good clothing at favorable prices. The owners had remodeled and updated many of the fixtures when the *setback* came—the building had been sold and they would have to move. After studying available buildings, the store management approached the owner of a former auto dealership two blocks away. The site was much larger than needed, so a portion was walled off and prepared for rental to the economy clothing enterprise.

The new location was convenient to the former location and the move took place smoothly. Business even increased in the new location since there was more parking available in that area of the city.

What the Cases Tell Us

- Some existing facilities can be satisfactory with modification.
- Some remodeling changes may suit the *structure* to a new use.
- A short move will likely be less of a business interruption than a longer move.
- Existing facilities may be more economical than a new building.
- All moves interrupt business.

BUSINESS FACILITIES

How Much Space is Needed by Your Business?

The size of a site may be as important as any other consideration in building an effective operation. An important question to consider is how much space is needed to merchandise or manufacture your product or service. One approach to sizing the enterprise would be to visit similar sites in the area and observe whether they seem adequate and well-arranged, or crowded and inefficient. Ideas for your office or facilities should be recorded for future use.

To use capital funds wisely, one needs to be prudent with the use of space. Adequate space should be allowed for efficient operation and customer convenience and approval. You all have probably had the experience of shopping in a store that was so crowded and cluttered that you felt hemmed in with no room to shop comfortably. On the other hand, you do not want the premises to appear empty or understocked.

Does Your Business Need Special Facilities?

To aid in locating and providing the needed facilities and special features, you need to list the essential considerations, such as ceiling hoist, traveling crane, or larger than normal doors, and those that would be helpful, but not required. From this, you can check off possible facilities to become aware of the best ones available, and determine their potential through modification or remodeling. Doors must be large enough for needed equipment, Figure 19–2.

What Utilities Will You Need?

Electricity is readily available in most communities of the United States. Why do we call your attention to it here? Some types of businesses require special *adaptations* of the utility, such as three- or four-phase current, and they are not always available. If you have special needs, working with the supplier can resolve the situation.

Water may or may not be available at your desired location, especially if it is outside the corporation or city limits. Many cities now create favorable rates for extending water and sewage lines to individual sites. This probably will not be a major concern for your enterprise at this time.

Sewage lines were mentioned above as they are often linked with the water supply. In locating a business, the capacity as well as the location of the sewage lines should be considered.

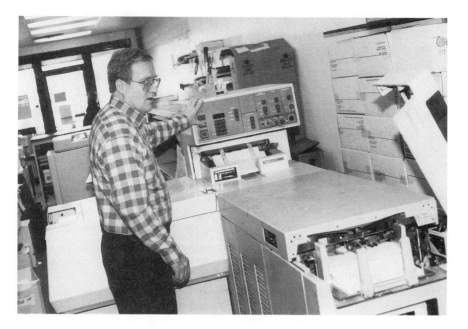

Figure 19-2 The outside door needed to be big enough to get this machine into Don's shop. (Courtesy Don Steele)

Should You Buy or Rent the Facilities?

Ownership allows complete freedom of choice in the use or changes of the facility, while it lessens the opportunity to seek a better location should the opportunity arise. Rental of facilities allows capital to be invested in stock and supply instead of facilities that limit cash flow. For businesses that are limited financially, renting or leasing offers significant advantages. Another advantage of renting is that as technology improves or conditions change, your business is not confined to an inadequate facility. Your business can move when the lease expires with more ease than if the facility were owned.

The same considerations apply to obtaining operational equipment as were discussed in terms of the facilities. Parking does not have to be considered in relationship to equipping the business as it is essential in considering site size. Figure 19–3 is an example of what to consider in deciding to buy or lease equipment.

SELF-EVALUATION

Facilities Needed Quiz

True-False: On a separate sheet, write T for true statements and F for false statements.

1. Equipment should always be purchased, never leased.
2. A list of required features is a good starting point for a search for suitable facilities.

3. Ownership and its freedom of control may be too high a price to pay for a beginning enterprise.
4. A large empty space may give the appearance of shopping comfort and convenience.
5. Landlords will often build or remodel a building to suit the tenant's need.
6. A mailing address and a telephone have launched many businesses.
7. A business moving from one location to another can do so without disruption.
8. Visiting similar businesses in other cities can answer your facilities problem.
9. Such a visit can give you some idea as to your size needs.
10. A short move rather than a long one can guarantee less business interruption.

STUDENT ACTIVITIES

1. Select a location and do a one-hour traffic count of passing cars. Compare your results with other students who observed other sites.
2. Count store traffic in a one-hour period in the same locations as before.
3. Compare traffic counts at a mall as compared to the city center.
4. Estimate space needed for a new business—the one you plan to start.
5. Compare leasing with buying computers, typewriters, and so on.

Figure 19-3   **Buy or rent? New or used? Your budget and desired image should help in decision making.**

CHAPTER 20

What Are the Costs of Running a Business?

OBJECTIVE

To identify the costs associated with a business enterprise.

COMPETENCIES TO BE DEVELOPED

After completing this chapter you will be able to:
1. List the common costs of running a business.
2. Identify variable and fixed costs.
3. Identify the relationship of production costs and level of production.
4. Identify the advantages and disadvantages of leasing versus purchasing machines.

TERMS TO KNOW

Administrative expenses

Depreciation

Direct materials

Fixed costs

Indirect materials

Investment tax credit

Legal expenses

Overhead

Renewal option

Variable costs

Cancellation penalty

Direct labor

Down payment

Indirect labor

Installment purchase

Lease

Obsolete equipment

Property tax

Settlement provisions

INTRODUCTION

One thing that successful business managers have in common is that they know the exact amount of their business costs. Your chances for success will be much greater if you have a good understanding of the expenses involved in starting and maintaining your business. In this chapter you will learn the answers to the following questions.

1. How do I figure the starting costs for my business?
2. How will I estimate my monthly expenses?
3. What are fixed costs? How are they related to my production?
4. What are variable costs? How are they related to my production?
5. What is a lease? Is leasing better than financing?
6. What provisions should be included in lease agreements?
7. What specific costs are entailed for my business?

CASE STUDY

Let's look at an example in which a good understanding of costs would have saved a new business manager not only money, but headaches as well.

The Case of the Critters in the Paint

Ed Johnson helped his grandparents paint their farmhouse one summer vacation. He decided that he not only liked the work, but was pretty good at it. When Ed graduated from high school the next year, he decided to go into business painting houses. His expenses wouldn't be great. He knew he would need ladders, brushes, and a paint sprayer. His father agreed to lend him the money to purchase the items. Ed placed an advertisement in the newspaper and a week later was called to make his first estimate.

He measured the dimensions of the house, subtracted the area of the windows and doors, and estimated that he would need about 12 gallons of paint. At $21.00 a gallon, the paint would cost $252.00. He also figured that if he and his grandfather had painted the farmhouse in 3 days, he could paint this house by himself in 5 days. He planned to make $8.00 an hour for labor, so he turned in an estimate of $575.00 for the job.

Ed's bid was accepted, and he purchased his supplies and paint for the house that afternoon. He agreed to start painting the next morning at 8:00.

The first day on the job went well for Ed. He finished one end of the house and trimmed around the windows on the front. He went home that evening feeling satisfied with his progress. The second day was much the same. Ed had the front of the house repainted and most of the garage trimmed. The third day, Ed finished the garage and began trimming on the back of the house. He discovered, however, that that end of the house had suffered more from exposure to the weather. He would have to spend time scraping old paint off the wood before he could apply the new paint. He spent the rest of his third day and most of the fourth scraping paint. He wasn't able to begin painting again until late afternoon of the fourth day. Because he was running behind schedule, Ed worked late that night, staying until after 8:00 when the mosquitoes and moths that were attracted to the outside lights finally drove him home. He left tired and discouraged. He would have to spend another 2 days painting in order to finish the job.

When Ed returned to work on the morning of the fifth day, he discovered to his dismay that the area he had painted the evening before was speckled with insects that had come out in the evening and had stuck to his wet paint. He returned to town to buy some steel wool and cleaned the insects from the side of the house. By the time he finished, his arms ached and he felt as though he could hardly lift his paint brush. Ed saw, however, that the area he had cleaned had a scratchy, dull appearance, and would have to be repainted. He spent the rest of the day repainting that section.

Ed returned the sixth day thinking he could finish the back of the house, but when he looked at his work from the previous day, he realized that the side with two coats of paint looked different from the side with only one coat. He would have to add a second coat to the rest of that end of the house to make it look even. That not only took more time, but extra paint as well. Ed's experience with the insects cost him an additional two gallons of paint. By the time he finished, he had spent almost 8 days painting.

The house Ed painted was 12 miles from his home. His daily drive, plus his trips into town for steel wool and extra paint, put over 200 miles on his pickup. Ed forgot to include pickup replacement costs, Figure 20-1. He spent $15.00 on gasoline; the extra two gallons of paint cost him $42.00; and the steel wool was $3.00. He spent an average of $4.00 a day for lunch at a nearby fast-food restaurant which totalled $32.00. Altogether, Ed spent $92.00 that he had not anticipated.

Ed had promised to pay his father $50.00 after each job until his debt was paid. After deducting from the total the cost of the paint, his payment to his father, and his extra expenses, Ed had only $178.00 remaining. Instead of 40 hours, the job took 67 hours. After calculating his time, Ed discovered that instead of making $8.00 an hour, he had only made $2.66 an hour, less than minimum wage.

Ed did have experience painting when he had finished his grandfather's house. What he didn't have was experience in figuring his expenses. He learned a valuable lesson and was thankful that he had only contracted to do one house before he discovered just how expensive those unexpected costs could be.

The owners of the house that Ed painted gave him a very good recommendation, and he painted several other houses that summer. But he never forgot the lesson he learned from his first job.

What the Case Study Tells Us

- Ed could have done a much better job of planning for his expenses. He learned the hard way that even small unexpected costs can add up.

- Ed should have taken a closer look at the job he was preparing to do. If he had foreseen the time he had to spend scraping paint, he could have adjusted his bid so that it included compensation for that extra time.

Figure 20-1 Depreciation was one cost Ed failed to include in his estimate.

- Ed did not think about the cost of his gasoline or the wear and tear on his truck. If he continued his painting business for several years, he would eventually have to replace his pickup. Unless he included that expense in his bids, the cost of replacement would have to come out of his profits.

BUSINESS COSTS

Hidden Costs

As Ed learned, there are often hidden costs in a business. Obvious costs are easy to see, but you need to think very carefully about other costs that may not be so apparent. The example we used was that of a service business, but there are hidden costs for every type of business. If you are manufacturing widgets, the expense of the materials and labor required to make each widget are fairly easy to calculate. Don't forget the cost of the money you borrowed to produce your widgets. That interest is just as much an expense as the paint you spray on each widget. So is the cost of manufacturing any faulty items that you can't sell, or the cost of fees that you pay for legal advice and advertising. When you are ready to borrow money, set your prices, and go into business, you will want to know the amount of every expense needed to make that business a success.

Starting Costs

There are many expenses that you will have to pay initially to get your business started. These are called starting costs. Some of these expenses, such as deposits for utilities, are only billed once. Others will come up again as you need to repair or replace equipment.

Fixtures and Equipment.   The cost of the fixtures and equipment that you will need to get started will depend on your type of business. A small office in your home will cost much less to equip than retail sales space for the public. Here are some of the items that you may need to purchase or lease.

- Counters
- Storage shelves
- Storage cabinets
- Display stands and shelves
- Cash register
- Safe
- Window display fixtures
- Special lighting
- Outside sign
- Delivery equipment

Don't forget that you may need to pay delivery and installation charges for some of these fixtures.

Decorating and Remodeling.   You may decide to keep your decorating scheme simple, especially in the early stages of your business. However, you do want to have a neat and clean atmosphere for your customers. Accomplishing this may be as easy as cleaning or repainting the facility that you have leased. You may, however, find that you need to do more expensive repairs, such as installing new flooring or lighting, before you can begin business.

Starting Inventory.   If you are running a retail or wholesale business, you will need to have your inventory established before you can open. Depending on your merchandise, this could be a major outlay.

Deposits with Public Utilities.   Many utility companies, such as telephone companies and electric services, require a deposit before they will begin service. In most cases, this will be refunded when you no longer use the service; in the meantime, the utility companies keep your deposit.

Legal and Other Professional Fees.   You may need to consult a lawyer about local ordinances or state and federal regulations concerning small businesses. Other professional fees might include the use of an advertising agency to promote your opening.

Licenses and Permits.   There are many regulations for the operation of a small business. These include various licenses and permits. Laws vary in different communities and states, but you can consult the local Chamber of Commerce, Area Planning Board, and Small Business Administration for information about the licenses and permits that you will need to obtain. You may find it helpful to talk with a lawyer who is familiar with the legal aspects of starting a new business.

Funds for Unexpected Expenses.   You will want to have some cash in reserve for unexpected expenses that may occur. You may need to make special purchases or you may have a loss or breakage. These could become additional costs.

Estimated Monthly Expenses

Once you have your business set up and ready to run, you will have monthly expenses. Some of these will vary throughout the year, but you will need to have a good idea of what to expect each month.

Your Salary.   Some entrepreneurs accept any surplus profit as their salary. Others prefer, or need, to put all profit beyond their minimum needs back into their company. Still others pay themselves a regular salary. You, as the owner-manager of your company, select the option best suited to your needs.

Other Salaries and Wages.   A salary is a fixed compensation for a service, usually computed weekly, biweekly, monthly, or annually. Wages are payments made on an hourly basis. An employee working for a wage usually keeps a time card which shows the number of hours worked each day. At the end of the pay period, the employee receives payment for the total number of hours worked during that time.

Monthly Property Payments.   Rent is a general term which covers the payments you make to use your business buildings. Actually, these agreements are often *leases*. If you purchase your buildings and property, you have mortgage payments to make. You won't have monthly rental or financing expenses, but you will have to pay *property taxes*.

Advertising.   This amount may be more difficult to estimate as you start your business. Your advertising budget may fluctuate quite a bit as you experiment with the best advertising for your situation. However, you should have an idea of the approximate amount that you will budget for your first year.

Delivery Expenses.   If delivery is a part of your service or sales business, you will want to budget for the expenses which you will incur. This would include the cost of a vehicle, taxes and license, wear and tear, and fuel, as well as the wage for the employee making the deliveries.

Supplies.   The cost of even small supplies can add up over a period of time. You will want to keep an ample amount of office supplies such as paper products, pencils, pens, markers, folders, staples, note pads, and other miscellaneous items. You will need to purchase invoices and sales receipts. You will also want to keep cleaning supplies on hand. If you have employees, you will need break room and rest room supplies.

Utilities.   These expenses are easier to estimate and will probably remain somewhat consistent unless your business is seasonal. They include the cost of telephone service, electricity, natural or bottled gas, and other utilities.

Insurance.   We discussed different types of insurance in the chapter about risk management. You will know what your monthly insurance costs are when you have purchased the insurance policies. For now, an insurance agent could give you an estimate of what your insurance costs will be.

Taxes.   You may need to talk to an accountant to get an idea of what your annual tax expenses will be. They will depend on your business income and expenses. You will have federal, state, and social security taxes to pay, and in some communities, a county tax.

Interest.   The interest you pay on the capital that you borrow for your business is a real cost. If you have a fixed interest rate, you will know exactly what your monthly interest expenses are. If your loan has a variable interest rate, you will need to estimate that cost.

Maintenance.   To run efficiently, machinery needs to have regular maintenance. Your business equipment needs this as well as your vehicles. Businesses frequently purchase maintenance agreements. This means that someone will come in at regular intervals to check your equipment for problems and to clean and oil your machinery. It usually costs less to keep machinery in good working order than it does to make expensive repairs resulting from a malfunction that has continued for too long.

You may also be responsible for the maintenance and upkeep on your buildings, even if you are renting or leasing them. (In that case, the party responsible for maintenance should be designated in the lease agreement.)

Legal and Other Professional Fees.   You may regularly use the services of a lawyer, accountant, or some other professional. If so, include this cost as you figure your monthly expenses.

Miscellaneous.   You must always plan for unexpected expenses. You may have to make unexpected repairs on machinery or purchases of equipment. It is wise to budget funds each month for unexpected costs which may appear.

Fixed and Variable Costs

The term "production volume" refers to the amount you produce. Your "sales volume" refers to the amount of your sales. If you are starting a service business, we could call the number of services that you render your "service volume." Your expenses will increase as your business grows. This is true for any type of business: wholesale or retail sales, service, manufacturing, or contracting. However, some of those costs will correspond directly with increases in your production. Others will remain somewhat consistent, even when production changes.

Fixed Costs.   While all of your costs will vary, some will not be dependent on your sales, production, or service volume. These expenses are called *fixed costs*. Some examples are your rent, your property tax, your *legal expenses*, and your property insurance. To a certain extent, your fixed costs will not change no matter what your production volume is. For instance, if you rent enough space and equipment to produce a maximum of 2,000 widgets, your rent will remain the same even if you only make 500 widgets. Here is a list of some of the fixed costs of running a business. Figure 20–2 illustrates some fixed costs in a production project.

- Building costs
- Storage costs
- Utilities
- Insurance
- Advertising
- Interest
- *Depreciation*
- Salaries and wages

- Transportation
- *Administrative expenses*
- Legal expenses

Figure 20–2   The feeders and the building represent fixed costs for Rob and his father. (Courtesy Kathy Shanks)

Of course, if you decide that you can sell 10,000 widgets a year, you will have to expand your facilities to produce that amount. You will have to hire more employees, rent more space, and buy more insurance. Your fixed costs will rise to a new level.

Figure 20–3 shows graphically how your fixed costs are affected by your production.

Variable Costs.   Some of your expenses will increase directly with your sales or production or service volume. If it takes 2 cents worth of paint to cover one widget, then it will take $20.00 to paint 1,000 widgets. If it takes $4.00 worth of gasoline to mow one lot, then it will cost $12.00 to mow three lots. These costs are called *variable costs.* Here is a list of variable costs for a business.

- *Direct material* costs
- *Indirect material* costs
- *Direct labor* costs
- *Indirect labor* costs
- Delivery expenses

- Bad debts
- Commissions
- All other overhead costs

Figure 20–3   Increases in fixed costs. (Courtesy The Three Entrepreneurs)

You will want to understand well the meanings of the above terms. Figure 20–4 illustrates a variable cost for Rob.

Direct Material Costs.   This is the cost of the parts and supplies that go into your manufacturing (or your jobs, if you have a service business). This also includes the cost of shipping, handling, and storage.

Indirect Material Costs.   These are materials too minor or unspecific to be included as direct materials, such as incidental supplies or machine lubricants.

Direct Labor Costs.   This is the cost of the labor to produce a product or perform a service. This would include the wages and salaries of those workers who actually manufacture or perform a service. *Indirect labor costs* would include clerical or janitorial workers.

Bad Debts.   Bad debts are those payments for your service or sales that you don't receive. Some businesses are more plagued by bad debt losses than others are. Unfortunately, legal action to recover bad debts is sometimes more costly than the loss itself. Some companies hire collection agencies to help them recover some of their bad debts.

Commissions.   A commission is a compensation, usually monetary, for sales. It is given as an incentive to employees for better sales performance. It is usually given in addition to a regular salary.

Other Overhead.   *Overhead* is a general term used to describe the expenses that go into a manufacturing, sales, or service business that are not included in your direct materials or labor costs.

The graph in Figure 20–5 shows how your variable costs increase with your production or service.

Not every business will have all the fixed and variable costs that we have listed. You may find some for your business that we have not included. The main idea is to know what costs affect your business, and whether or not those costs are fixed or variable.

Figure 20–4   Feed for the sows represents a variable cost for Rob. (Courtesy Kathy Shanks)

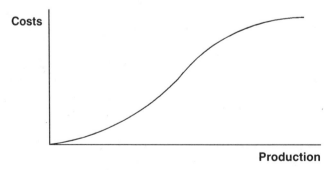

Figure 20–5   Increases in variable costs. (Courtesy The Three Entrepreneurs)

Should I Lease or Finance?

One way many businesses reduce their costs is by leasing or renting their buildings or equipment. A lease is similar to a rental agreement. However, a lease is usually written for a longer period of time, and has more specific provisions in the contract.

You will be able to choose whether to lease or buy several of the assets your business will use. The decision to lease or buy buildings will affect the location that you select. If you move your business into an existing building, you will probably lease the space that you use. If, however, you purchase a lot at the edge of town on which to establish your business, you will pay for the buildings you erect on that space. The alternative would be to talk an investor into purchasing the property and building the facilities, and then leasing them to you.

Leasing can apply to more than the property and the building that houses your business. You may also want to consider leasing your production or service equipment and machinery. A building contractor may need to use a backhoe only two or three times a year. It might save money to lease that piece of machinery for a short time rather than to purchase it. The costs of leasing equipment versus purchasing it need to be compared.

Few entrepreneurs have the capital needed to get a new business off the ground. This means that they borrow money to get started. In that case, a lease can save money that otherwise would have to be borrowed.

There are several possible sources from which you can obtain a lease. These companies are risktakers, just as you are, and are in the leasing business in order to make a profit. Therefore, they will make money from you. There are companies that specialize in leasing buildings or equipment. In addition, commercial banks, insurance companies, and finance companies often lease land, buildings, or machinery. Before signing a lease agreement with a little-known company, check with the Better Business Bureau in order to be sure that you are dealing with a reliable company.

Lease Agreements.   The lessee is the person (or company) who uses the asset. The lessor is the person (or company) who owns the asset. There are several items which should be specified in a lease agreement.

1. The financing arrangements and the payment amounts.
2. The term of the lease.
3. The *settlement provisions* in case of damage or destruction.
4. The party that receives the investment credit.
5. The party that is responsible for maintenance and taxes.
6. The *renewal options* that exist.
7. The *cancellation penalties*.
8. The party that will make insurance payments for the leased equipment or buildings.
9. Any special provisions.

There are advantages and disadvantages to leasing. We have made a list of some of the most important ones.

Advantages of Leasing

1. *No down payment.* A new business will need a *down payment* of about 25% to borrow money for a major purchase. When you lease, that's capital that you can put into other operating expenses.

2. *No financial restrictions.* When you borrow money from a lending institution, it will often closely watch your operating expenses, and may even place restrictions on your expenditures. When you lease, you retain control of your finances.

3. *Longer period of payments.* A lease is usually set up for a longer period of time than a loan. That means that you spread your expenses over a greater number of years, and pay less per year. This is especially beneficial to the entrepreneur who has less operating capital in the first few years.

4. *Tax benefits.* A lease is a legal operating expense and is therefore tax deductible. However, for a lease to qualify as a business deduction, it must be a true lease and not an *installment purchase.*

5. *Disposal of obsolete equipment.* When the duration of the lease has ended, the asset is returned to the owner. A lease makes it easy to update or replace your equipment, or even to move your business, at the end of the lease term.

6. *Short-term use of equipment.* If you need a special piece of machinery for a limited time, you might consider leasing it for several months rather than purchasing and reselling it. This would be beneficial for a special project or for a seasonal business.

7. *Known payments.* Payments for a lease are established at the time the contract is signed. These payments remain constant throughout the life of the lease. If money is borrowed at a variable rate for a purchase, the payments will fluctuate with the changes in the interest rate.

8. *The use of expensive equipment.* A lease may allow you to use equipment that you otherwise would not be able to afford.

Disadvantages of Leasing

1. *Binding agreement.* A lease is a legal obligation to continue payments until the end of the term. That commitment remains even if you no longer want or need the equipment that you have agreed to lease.

2. *Loss of asset.* When a lease expires, the asset returns to the owner. Certain assets appreciate over time. The lessee loses this advantage. When you purchase an asset, it belongs to you after the final payment has been made.

3. *Loss of investment tax credit.* The owner of an asset receives the *investment tax credit* for the purchase. This can be a substantial sum for a large purchase.

4. *Lack of convenience.* Remember our example of the building contractor who needed a backhoe only occasionally. While it might cost less for him to lease the machine, he would want to be sure of its availability when he needed it. If he had to pay workers while he waited for the use of the backhoe, he would have been better off to buy it himself.

Subcontracting: Another Alternative

Some business managers prefer to subcontract certain jobs rather than to lease or purchase equipment and hire the labor to do it. This method can be especially helpful when the service or job is one that is not directly related to your normal business or one that is not performed frequently. Subcontracting may or may not cost less than leasing. It depends on the situation and how often you would need to subcontract a particular type of job.

There are, however, things that you want to be aware of if you subcontract work. Most important, you want to be sure that you are dealing with a reliable company. Can they finish the work according to your time schedule? Will the quality of their work be consistent with your standards? You wouldn't want your business reputation harmed because of someone else's poor work.

SUMMARY

Business costs vary greatly according to the type of business. The expenses of a service or contracting business may be very different from those of a manufacturing, retail, or wholesale business. Your student activities manual has a list of expenses for several different types of businesses. Select the one that is closest to the type of business that you are starting. You can fill in approximate expenses for your business.

SELF-EVALUATION

Business Costs Quiz

1. List the common costs of running a business.
2. Identify the following items as fixed or variable costs by writing an F for fixed costs or a V for variable costs.
   - a. Raw materials
   - b. Employee salaries
   - c. Employee wages
   - d. Equipment purchase
   - e. Delivery costs
   - f. Design costs
   - g. Utilities
   - h. Machinery maintenance
   - i. Building maintenance
   - j. Leasing charges
3. What are the differences between variable and fixed costs?
4. If fixed costs total $1,000 and variable costs are $10 per unit, what is the cost per unit at 100 units? at 200 units? at 500 units? at 1000 units?
5. List the advantages of leasing.
6. List the disadvantages of leasing.

STUDENT ACTIVITIES

1. List the costs on your business plan.
2. Identify these costs as fixed and variable costs.

3. Set up a hypothetical production project and list the costs associated with the project.
   For example: a. a sow and litter project
                 b. a popcorn production project
                 c. a lawn mowing business, etc.

CHAPTER 21

How Do I Price My Product or Service Fairly?

OBJECTIVE

To identify pricing methods.

COMPETENCIES TO BE DEVELOPED

After completing this chapter you will be able to:

1. Compute prices given the wholesale price and the percentage of markup.
2. Compute the cost per unit given the number of units and fixed costs.
3. Correctly identify the highest profit point of production given the projected sales at given prices and the cost of production per unit.
4. Identify the items that contribute to the pricing equation.
5. Identify ways of forecasting sales.

TERMS TO KNOW

Break-even	Discount
Fixed cost	Forecasting
Markdown	Markup
Price	Profit
Sales quota	Sales volume
Variable cost	Widget

INTRODUCTION

Pricing is a challenging concept because there are no exact answers for you as a business manager. A new pricing situation is created with each new business. Yet your prices will have as much to do with the success or failure of your business as any other single factor. There are many different things that you need to take into account as you decide how to set your prices. Consider the following:

- How do my *fixed costs* influence my prices?
- How do my *variable costs* affect my prices?
- How will my prices influence my *sales volume?*
- How do I know when I'll break even?
- What is a *sales quota*?
- How do I know what my customers will be willing to pay?
- How will my pricing strategy affect my business image?
- How will my pricing strategy influence the way customers perceive my product or service?
- How can I experiment with different prices?
- When are prices sensitive to change?
- Can I underprice the competition?
- How can I figure *markup* percentages?
- How will I establish my prices?

CASE STUDY

Phil Learns the Facts About Pricing

Shortly after Phil took over the management of the elevator portion of the feed and grain business, he emptied the corn bin to see if he could account for the number of bushels purchased and put into the bin in the first place. Imagine his surprise when he found that he had put many more bushels into the bin than he had sold!

Phil had not considered the shrinkage as grain loses moisture and the losses due to spillage or pests. This amounted to approximately 5% and was considered too high.

What the Case Tells Us

- There are losses in handling any commodity.
- Losses can be determined and controlled.
- These losses must be considered in setting the prices.

WHAT FACTORS INFLUENCE MY PRICING?

When setting your prices, remember that you want to recover all of your costs and make a profit, while selling at a price that the public is willing to pay. You must know exactly what your costs are. If you sell below cost, you will lose money. Any sales above cost are your profits.

If you price your product or service too high, your sales will decrease, but you can also lose sales by pricing too low. That's because your prices affect the way the public perceives your product. Higher prices frequently imply better quality to the consumer. This is especially true for a business that offers a service. You want to set your price at the point where you can sell enough of your product or service to recover your costs and still make the best possible profit. Renita likes horses so her business choice was easy, Figure 21-1.

Figure 21-1   Renita's boarders enjoy the same care she gives her own horses. She must set a price that provides a fair return for facilities, labor and feed. (Courtesy Lester Bean)

How Will My Prices Influence My Sales Volume?

In the preceding chapter, we discussed the different costs of running a business. Remember that your variable costs, the cost of materials and labor, increase with each unit you produce. Your fixed costs, the cost of your rent, and so on, remain the same within a certain range of production. As you increase your production within that range, your cost for each unit decreases.

Let's say that you start a manufacturing business and set up equipment and space (your fixed costs) to produce 2,000 widgets a year. If you only sell 500 widgets a year, then the price you receive for each *widget* will have to recover 1/500 of your fixed costs. If you can sell 2,000 widgets a year, then each widget will need to recover only 1/2,000 of your fixed costs. These fixed costs per unit of production will decrease as your sales volume increases. When your cost per unit decreases, you have two options. You can lower your prices or increase your profits. That brings about another decision. If you lower your prices, then you'll want to produce more than 2,000 widgets because the demand will be greater for your less expensive products. That means an increase in your production capacity. Your fixed costs will increase to the next level, and those added costs will need to be included in your new prices. You can begin to see how challenging (and important) your pricing decisions will be.

There is a direct relationship between the *price* that you set for your product or service and the amount of that product or service that you sell. The highest price that you can receive may not create the best business results. Conversely, your sales volume may increase as you lower your price. As a simple illustration, let's say that the variable cost of an item you are producing is $3.00; that is, you need exactly $3.00 worth of material and labor to make each unit. Your are considering pricing each unit at either $4.00 or $5.00. Suppose that at $5.00 each, you estimate that you can sell 10,000 units but at $4.00 each, you think you can sell 30,000 units. The amount of money you make over your variable costs is the amount that you can apply to your fixed costs and a *profit*. Figure 21–2 illustrates what happens.

PRICING OPTIONS

	$5	$4
Selling price	$5	$4
Number of sales	10,000	30,000
Dollars sold	$50,000	$120,000
Variable cost	$30,000	$90,000
Dollars left to pay fixed costs and profit	$20,000	$30,000

Figure 21–2 Price versus sales. (Courtesy The Three Entrepreneurs)

You can see that $4.00, the lower price, would be the best price in this case. However, if you are working with a limited market which will purchase only 10,000 units during a given period, but which is willing to pay $5.00 for each unit, then $5.00 would be the best price.

Break-Even Analysis

Your *break-even* point is that level of sales where your costs equal exactly what you are taking in. At the break-even point, there is no profit and no loss. It's important to be able to calculate your exact break-even point. Above that level you are generating a profit. If your sales are below your break-even level, it's simply a matter of time before your business is bankrupt.

There is a simple formula that you can use to calculate the amount of sales you need to break even. Your break-even point is equal to the sum of your fixed costs and your variable costs:

Sales at Break-Even = Fixed Costs + Variable Costs

Remember the graphs of fixed costs and variable costs from the preceding chapter? We can combine those to show a total cost line. Then we can plot a total revenue line to show the break-even point. This gives us a simple illustration of the relationship of profit and loss to costs and revenue. Our graph shows a hypothetical situation. When you know the exact fixed and variable costs of running your business, you will be able to chart your projected revenues at various prices to find the break-even point for your business. You have already made a balance sheet and you can use the costs that are detailed on the expense list. You have also projected a cash flow for a period of time. Your cash flow will depend directly on your sales volume and your prices, but your projected cash flow will help you determine the level of your prices. Figure 21–3 will illustrate.

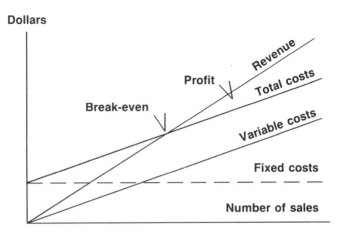

Figure 21–3   Break-even analysis.

Let's look at an example of break-even analysis. Rosie Fern wants to open a flower shop and needs to know what her break-even point would be. She has figured her fixed cost for the first year at $9,000. She also figures that her variable cost will be about $700 for every $1,000 of sales, or 70%. She wants to know what her sales must be the first year in order to cover her costs. Using the formula above, she knows that her sales must equal her fixed costs plus her variable costs, or $9,000 plus 70% of her sales.

$$Sales = \$9,000 + .70 \ (Sales)$$

or

$$S = \$9,000 + .70S$$

Using simple algebra we can figure Rosie's break-even point.

$$S = \$9,000 + .70S$$
$$S - .70S = \$9,000$$
$$.30S = \$9,000$$
$$S = \$30,000$$

Rosie will need to make $30,000 her first year to break even.

Adding in a Profit

At first glance it looks like anything that Rosie makes over her $30,000 will be profit. Remember that there is a cost included with each sale—as her sales increase, so do her expenses. Now Rosie would like to know what her sales must be to give her a $12,000 profit. We can include that in our formula.

$$S = \$9,000 + .70S + \$12,000$$
$$S - .70S = \$9,000 + \$12,000$$
$$.30S = \$21,000$$
$$S = \$70,000$$

Rosie has to more than double her sales in order to make a $12,000 profit.

Now let's change Rosie's situation somewhat. She had figured her variable costs at $700 for every $1,000 worth of sales. Let's double her prices and see what happens. Now her variable costs will be $700 for every $2,000 worth of sales, or 35% instead of 70%. Using the same formula (Sales at Break-Even = Fixed Costs + Variable Costs), where is Rosie's new break-even point?

$$\text{Sales} = \$9,000 + .35 \text{ (Sales)}$$

or

$$S = \$9,000 + .35S$$
$$S - .35S = \$9,000$$
$$.65S = \$9,000$$
$$S = \$13,846.15$$

Rosie's break-even point at the higher prices is less than $14,000 instead of $30,000.

Now let's see what Rosie's level of sales must be in order for her to make a $12,000 profit.

$$S = \$9,000 + .35S + \$12,000$$
$$S - .35S = \$9,000 + \$12,000$$
$$.65S = \$21,000$$
$$S = \$32,307.69$$

You can see that if Rosie doubles her prices, she'll break even and make a profit much sooner than at the lower price.

This illustration is to show you the difference your prices can make in your business. It's unrealistic to think that Rosie could get away with doubling her prices. Perhaps she could, but her sales volume would be cut in half. It's possible that Rosie could have $70,000 worth of sales in a year's time. At the higher price, maybe it would take a year for Rosie to sell $32,307.69 worth of her flowers. Either way, she'd be making a $12,000 profit that year. There may be some price in between the highest and the lowest at which Rosie could keep her sales volume high and make an even better profit. These are some of the pricing decisions that Rosie must make.

Details can make a difference in the profit picture. Dan-D Acres again is an illustration. When strawberries were picked in the morning they weighed more than when picked in the afternoon, Figure 21–4.

Figure 21-4 Berries picked in the morning weighed more than those picked in the afternoon, thus affecting returns. (Courtesy Barbara Doster)

Forecasting Sales

Before you can decide your production level or set your prices, you need to know about what to expect for your sales. We could compare sales *forecasting* to weather forecasting. You, like the meteorologist, will be making an educated guess using all the data that you can gather. Here is a list of some important questions to help you forecast your sales.

- Who will buy my product or service?
- How much will they pay?
- How often will they buy?
- In what amounts will they buy?
- Where are they located?
- How far will they come to buy my product or service?
- If I have a manufacturing or sales business, what sizes, colors, or styles will my customers prefer?
- If I have a service business, what times of the year will people pay for my service? How often?
- What is the sales performance of my competitors?
- What are the general trends in similar businesses?

There are several ways to find the answers to these questions. You can talk to salespeople in similar businesses. You might talk with someone running a business like yours in another community or send questionnaires through the mail. (You might offer a coupon in exchange for their response.) Some market researchers use methods such as street corner surveys or telephone polls. You want to establish as closely as possible the total sales potential for your sales zone. By studying your current competition, you can estimate your sales potential.

Good marketing research is expensive. However, poor research can be more costly when it leads to wrong decisions. Fortunately, there are many agencies that are set up to help small businesses with this sort of problem.

Can I Underprice My Competition?

The answer to this question depends on your competition. It is unusual for a small business to successfully compete with prices set by a large company. Big businesses have usually operated long enough, and at a large enough capacity, to reduce their costs to the minimum level for their production. They buy their materials in bulk at low prices. They often run at an efficiency that a new business finds hard to match. If you find your business competing in this situation, you'll need to give consumers a reason to pay your prices. This is the time to emphasize the convenience or quality of your product or service.

If you are competing with a similar-sized business in your community, then it's possible that you can underprice your competition. You may want to accept a lower profit as you establish your business, especially when trying to penetrate a market that has previously belonged to another business.

How Do I Set a Sales Quota?

A sales quota is a goal that you set for your business. Once you have forecasted your sales, you can establish an objective sales quota that will tell you how to prepare for that level of sales. It's important to be as accurate as possible about anticipated sales. If you are too optimistic, your production costs will be higher than your sales. If your projection is too low, you will lose potential business. Your actual sales may differ from your anticipated sales, but the activities and resources (your fixed costs) that will support that level of sales will not.

Pricing Considerations

Here are some pricing considerations that you may want to keep in mind as you establish your prices.

1. *Odd-ending prices.* There is research supporting the belief that consumers feel they are getting more for their money when a product costs less than the next whole dollar. There seems to be a psychological advantage to pricing a product at $9.99, for instance, rather than at $10.00.
2. *Multiple pricing.* Some businesses offer *discount* volume purchases. For example, you may sell one unit of your product or service at $15.00 or two at $28.00.
3. *Leader pricing.* Retail stores often advertise an item at cost, or less, in order to bring customers into the store. Research shows that customers frequently make additional purchases once they are drawn into a store.

4. *Wholesale pricing*. Wholesalers deal with large quantity sales to a select group of customers (retailers). Retailers then add on their markup before they sell to the general public.

5. *Rate of turnover*. Items that move slowly are often priced higher than items that sell quickly. Don't forget that it costs something to store an item until it sells.

6. *Installation expenses*. If you are selling a product that must be installed, you'll want to include your installation expenses in your price or charge an additional installation fee.

7. *Markdowns*. Markdowns are reductions in prices. Items may be marked down either to bring in customers, or to move the items out of inventory. For instance, rather than storing a seasonal item, many businesses will lower the price at the end of the season.

8. *Spoilage/breakage*. Items that spoil or are broken are a direct loss to a business. If possible, you should adjust your price so that the units that you sell cover the expense of those that you can't sell because of spoilage or breakage. Strawberries, for example, have a short shelf life, Figure 21–5.

Figure 21–5   Price sensitivity is illustrated by strawberries, which spoil quickly. (Courtesy Sadie Hudgins)

9. *Delivery*. Some businesses advertise free local delivery. Like free installation, the cost of free delivery is included in the price of the item.

10. *Employee discounts*. Many businesses offer employee discounts as a company benefit. This, too, should be taken into consideration as you develop your pricing strategy.

Pricing Regulations

Many states have laws which regulate sales practices. These are often referred to as unfair sales practices acts. These regulations may establish a minimum markup on certain products. You'll want to know what regulations exist in all the states in which you'll be selling your product or service.

Price Sensitivity

In some situations the public may be very sensitive to the prices that you set. Even a few cents can send your customers to a competitor. In other situations, you may be able to increase your price substantially without chasing customers away. Here are some of the factors that influence price sensitivity.

Supply.   The supply of a product or service is the amount that is available. If the supply of a desired product is limited, customers will be willing to pay a higher price for it. If there is an unlimited supply, you probably want to look for another product to market. For example, a late frost in Florida reduces the orange crop and the price of orange juice goes up. When farmers have a bumper crop, the supply goes up and the price goes down.

Demand.   As the desire for a product or service increases, its price increases. This economic concept is called the level of demand. There are more swimming pools and air conditioners sold in the summer months. People will pay more for a snow shovel when a January blizzard is approaching than they will in the heat of July.

Availability of Substitutes.   If you are the only business in a small town that paints houses, you will be able to charge more for your services than if a similar business exists in your community. If you have a product that cannot easily be replaced, you may be able to charge higher prices. However, if the consumer is easily able to substitute another product for yours, you'll find that in order to remain competitive, you'll need to sell competitively, or give your customers a reason to pay your higher price.

Frequency of Purchases.   Consumers are willing to spend more for items that they purchase only occasionally. If your product or service is one that a consumer buys regularly or frequently, he or she is more apt to shop around for the lowest price.

Buyer's Budget.   We have mentioned the importance of knowing your market. One of the most critical factors that influences a consumer's sensitivity to prices is his or her own budget. If your product or service is designed for a group of consumers that has a lot of excess cash, its price will be less critical than if you are selling to a group of consumers on tight budgets.

Establishing Prices

There are several different methods that businesses use to set their prices. Here are a few of the most common.

Pricing by Cost.   This method is the one most frequently used by small businesses. After a thorough analysis of the costs that go into your product or service, add your desired profit to establish your price. This formula is simple.

Direct Labor + Direct Materials + A Percentage of Overhead
+ Your Desired Profit = Your Price

As a simplified example, let's say that you are making doghouses and have orders from 10 customers. You can make one doghouse in 4 hours. At $5.00 an hour, you figure your direct labor at $20.00. The wood, nails, shingles, and paint for each doghouse cost $10.00. You purchase a circular saw and a hammer for $185.00. You estimate that you'll use $15.00 worth of electricity, so your total overhead is $200.00. Divided among 10 doghouses, that's $20.00 each. Your total cost for each doghouse is $50.00. Now, assume that you want a 15% profit. Your profit (remember that your labor is a cost) will be $7.50 for each doghouse. You'll need to set your price at $57.50.

$$\$20.00 \; + \; \$10.00 \; + \; \$20.00 \; + \; \$7.50 \; = \; \$57.50$$

This would be the first level price for 10 doghouses. Now, if you contract with a lumber company to sell your product, you become the wholesaler and the lumber company becomes the retailer. The lumber company will mark up your price. If you anticipate selling more than 10 doghouses through the lumber company, you can reduce your first level price because your fixed costs will be spread out more.

Competitive Pricing.   Competitive pricing is generally used for items or services which are common and are hard to sell at other than the established market price. This method is commonly used by oil and steel producers and auto makers. When you set your prices according to competition, your first priority is to see that your costs are covered and that you still make a profit.

Market Pricing.   Using this method, you set your prices according to what the market will bear. In other words, you price your product or service at the highest price that the consumer will pay. This method of pricing is every business person's dream, but it's the least commonly used. It works best for unique items that have a limited availability. Market pricing seldom lasts long because the desire to share the profit will create competition.

Changing Your Prices.   Even the best market research cannot tell you exactly how to price your product or service. Pricing always includes some experimenting on the part of the seller. What do you do when the public won't accept the prices that you have set for your product or service? Here are some things that you should keep in mind.

1. *Public acceptance of price changes.* It's easier to lower your prices than it is to raise them. You may want to establish a price scale ranging from your most optimistic price to your minimum profitable price. Try selling at your optimum price. If you can't maintain the sales volume you need at that price, then you can lower it. If you begin pricing at the lower end of the range, you may alienate some of your loyal customers when you find you need to increase your prices.

2. *Accept a lower margin.* You may be forced to accept a lower margin of profit. Your goal may be a 20% profit on your sales. If you don't have enough customers at that price level, you will need to reduce your prices and accept less profit.

3. *Reduce costs.* Another alternative is to reduce your costs. There may be a more efficient way to produce your product, or a less expensive way to offer your service. Successful business managers find ways to cut some of the frills out of their businesses.

4. *Differentiate your product or service.* You may be able to beat the competition by emphasizing quality, service, performance, delivery time, or financing arrangements. Consumers will pay a higher price for what they see as a better buy.

5. *Discontinue a product.* You may find that no one will buy a certain product at the price you need to charge. If all other alternatives fail, it's best to simply discontinue that product, perhaps replacing it with another.

Pricing for Different Types of Businesses

Pricing strategies vary for different types of businesses. While the basic concept of pricing remains the same—each business must recover its fixed and variable costs and make a profit— different factors affect the prices that various types of businesses can charge. You are now in the planning stages for a specific type of business.

Service.   The prices for a service type of business are usually the easiest to establish. Your investment in the business includes your skill and knowledge plus the equipment you need to perform the service. Your fixed costs are equipment, advertising, and insurance. The additional expenses that are involved each time you perform the service (such as gasoline, seed, and chemicals for a lawn care service) are the variable costs. You can place a value on your time and knowledge (which will depend on your competition and the demand for your service) and add on your costs to establish your prices.

Sales.   Pricing for a sales business depends more on the existing competition. Customers will pay a higher price at your business if they must drive 20 miles to purchase the item at a lower price. The level of your prices will be more critical if less expensive substitutes are readily available.

Manufacturing.   The prices set by manufacturers are among those most dependent on costs, unless the manufacturer is producing a unique product.

Contracting.   A contracting business compiles the resources of other businesses to complete a job. Those other businesses are subcontracted. A contractor's prices will often fluctuate according to the subcontracting costs.

Your Pricing Strategy

As you can see from the discussion in this chapter, pricing decisions are highly individual. The prices that you can set for your products or service will depend greatly on your business costs. They will also depend on the type of product or service you offer, the people you are selling to, what they are willing to pay, and the nature of your competition. With careful planning, your pricing strategy will contribute to the success of your new business venture.

SELF-EVALUATION

Pricing Quiz

True-False: On a separate sheet, write T for true statements and F for false statements.
 1. Prices will have a minor effect on the success or failure of a business.
 2. Higher prices are synonymous with higher quality.
 3. Prices do influence sales volume.
 4. At the break-even point all fixed and variable costs are covered.

5. At the break-even point, management return is provided.

6. Doubling sales will generally double profits.

7. A person can forecast sales with a high degree of accuracy.

8. Underpricing competitors is one method of penetrating a market.

9. Research does not support the odd-ending pricing pattern that is so common.

10. Discounts for volume purchases are unfair to small buyers.

11. Leader pricing is often a good practice.

12. Wholesale prices are always lower than retail prices.

13. Slow moving items may require a higher markup than fast moving items.

14. Markup means the increase in selling price over purchase price.

15. Sales quotas are objectives.

16. Employee discounts are fringe benefits to an employee.

17. Supply and demand play an important role in pricing.

18. It is more acceptable to raise your prices than to lower them.

19. Lowering margins and reducing costs are two alternatives to raising prices.

20. Differing types of businesses must use differing pricing strategies.

Production Level

Given the following facts, determine how many widget machines will be produced at the following costs and returns. Tabulate your calculations on a separate sheet.

NUMBER OF UNITS	FIXED COSTS	VARIABLE COSTS	TOTAL COST	PRICE	TOTAL RETURN	PROFIT OR LOSS
10						
20						
30						
40						
50						
60						
70						
80						
90						
100						

Fixed costs are $2,000. Variable cost per unit is $250.

Prices at:	1–30 units	$350
	31–50 units	350
	51–70 units	325
	71–100 units	290

STUDENT ACTIVITIES

1. You are manufacturing a product known as Willy's Widgets. Your fixed costs for the month total $3,000, and the cost of producing each widget is $10.00. If you can sell a widget for $16.00, how many widgets must you make and sell per month to break even?

2. If 500 units are needed each month, how many units must be sold to make $1,000 per month profit? to make $2,000 per month profit?

3. If 1,000 units per month would reduce the variable cost per unit to $7.50, but the price would need be $14 per unit, which would be the best production level: 500, 600, 700, 800, or 1,000? If the price were to drop to $12 at 1,500 units and the variable costs stay at $7.50, how many would the business produce?

4. In its first six quarters, the sales of a new business increased as follows:

quarter 1	$3,000
quarter 2	4,000
quarter 3	6,000
quarter 4	8,000
quarter 5	11,000
quarter 6	13,500

What would be a reasonable forecast for quarters 7, 8, 9, and 10? Discuss the answers.

CHAPTER 22

How Do I Schedule My Production, Service, or Sales Activities?

OBJECTIVE

To explore the concepts of scheduling production or services.

COMPETENCIES TO BE DEVELOPED

After completing this chapter you will be able to:
1. Explain the concept of business scheduling.
2. Explain the cash flow cycle.
3. Describe the problem-solving method.

TERMS TO KNOW

Adept	Avert
Cash cycle	Cash flow sheet
Credit	Credit agreement
Credit bureau	Finance charge
Logical steps	Meticulous
Scheduling	Small Business Administration

INTRODUCTION

To an inexperienced person, the concept of scheduling business activities might seem simple. You buy materials, produce a product or service, and sell it, right? You know by now that the profitable management of any business involves more than that. *Scheduling* means careful planning for all aspects of operating a business. As a business manager, you will be doing a lot of planning—both short-range and long-range.

Important topics in this chapter deal with planning. The following is a list of questions to keep in mind as you think about scheduling for your business:

- What does the term "cash flow" mean?
- How will my inventory affect my cash flow?
- How can I *avert* cash flow problems?
- When will I need to borrow capital to facilitate my cash flow?
- Should I extend credit to my customers?
- How will I schedule my production activities?
- How can I fix production mistakes?
- How can I learn problem-solving skills?

CASE STUDY

The Case of a Small Beginning

Jerry Hawkins lived with his parents in the suburbs of a metropolitan area. One day, while rearranging their garage, Jerry and his father decided they had accumulated so many things that soon they would soon have no room for the car. They decided to build a small storage shed for bicycles, the lawn mower, rakes and shovels, and the backyard grill. They spent about $400 on the shed, which they built in two weekends. When they were finished, they had an attractive building that solved their storage problem. Both Jerry's uncle and a neighbor were interested in the shed, and asked Jerry and his father to build similar sheds for them. Jerry's father felt that he was too busy to devote more weekends to carpentry, but he agreed to oversee the projects if Jerry and his cousin, Ned Riley, wanted to build sheds together to earn spending money. The jobs went smoothly. Jerry and Ned turned out two attractive, well-built storage sheds. The project gets started, Figure 22–1.

Jerry and Ned now have a new skill. They built the third shed in about half the time that the first one took. They believed that other people would commission the building of similar sheds, and they advertised their product. In order to convince customers to hire them, they offered a guarantee of satisfaction. They contracted to build an additional six storage sheds, with no charge to the customer until the product was delivered. Jerry's father agreed to finance the materials, with the agreement that the two boys would pay him a finance charge for the use of the money.

Because the boys were out of school for the summer, they could work steadily at the project. They were building six sheds at once, so they could develop a small-scale assembly line, which improved the efficiency of the project. The sheds were finished in three weeks. All six customers were satisfied with the quality of work that had been done. The sheds were delivered and Jerry and Ned received payment promptly. They were able to repay Jerry's father and keep a good profit for themselves.

Figure 22–1 Laying out a skid is the first step in building a storage shed. (Courtesy Donald F. Connelly)

This is a simplified version of the inception of Hawkins and Riley Construction Company. Jerry and Ned remained in business together after they finished school, expanding their product from storage sheds to pole barns. The company now has seven additional employees. Jerry spends most of his time managing construction and overseeing work at the job sites, while Ned handles marketing and sales.

The two men agree that they learned some valuable business lessons from those eight storage sheds that they built several years ago. One of the most important lessons was about the impact of cash flow on financial management. Their business operates in much the same way today as it did when they were building as teenagers, but on a larger scale. Their cash outlay is extensive, and so is their cash intake. The timing is not always the same, so that sometimes they are spending more money than they are taking in, while at other times, they are taking in more money than they are spending. Sometimes they are borrowing capital to pay for materials. At other times they are banking money so that it will earn interest.

What the Case Tells Us

- Most small businesses start very small.
- Many businesses grow from a money-making project that succeeds.
- Payment on completion complicates cash flow. Most construction operates on a pre-determined partial payment basis.
- *Credit* is a necessary adjunct to business activity.
- Volume provides efficiency.
- Early success builds self-confidence.
- Satisfied customers build future business.
- Guaranteed work attracts customers.
- Support is needed for start-up.
- Mass assembly and quality control grew from experience in the assembly process.
- Partners pooled resources and divided responsibilities.
- It is important for partners to work well together.
- Cash flow is necessary regardless of the size of the business.
- Inventory ties up capital.

CASH FLOW

What is Cash Flow?

As discussed earlier, volume of sales is important to the success of your business. What happens when payments for sales or services aren't received immediately? There will still be bills to pay, and your employees will expect the payroll to come through. You may have additional investments planned. Will you have sufficient cash to keep your business in operation during the times that money is not coming in?

On the extreme, what will you do with large amounts of incoming cash when you expect no major expenditures? Pocket it and take a vacation? No—you want to put it to work for your business.

Put simply, cash flow is the difference between what you take in and what you spend during a given period of time. A *cash flow sheet* is a valuable management tool. With proper planning, you will be prepared for large expenditures. You will also be able to plan to make large amounts of cash work for you.

Every business has a *cash cycle*. Cash comes in, and cash goes out, but not necessarily at the same time or at the same rates. The diagram in Figure 22–2 illustrates the cash cycle of a business.

You can chart the incoming and outgoing cash, or capital, for various periods of time. You can create an actual cash flow sheet, which charts the flow of cash in the past, and then use this tool to plan for your future, or projected, cash flow.

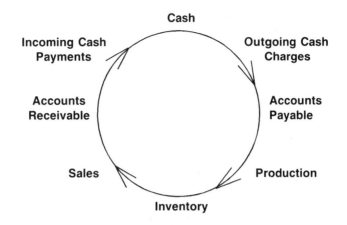

Figure 22–2   Cash flow. (Courtesy The Three Entrepreneurs)

How Does Inventory Affect Cash Flow?

In Figure 22–3, we illustrate the position of the inventory in the cash flow cycle or circle.

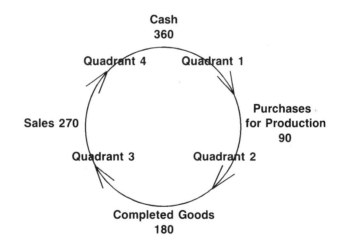

Figure 22–3   The cash flow circle. (Courtesy The Three Entrepreneurs)

In the first quadrant, cash is flowing to purchase of materials for production. In the second quadrant, it provides for wages for the worker, factory overhead, and other costs that go into making the complete goods. In the third quadrant, we are 180° away from the cash position. Now, sales start to bring in cash, returning toward the cash position. The more of our resources that are tied up in inventory, the greater the need for supplemental cash to keep the business going. This can be obtained by additional capital inputs or by borrowing from lenders.

How Can I Avert Cash Flow Problems?

Management and coordination are the means of controlling the cash flow circle. The completion of goods that can be shipped instead of stored illustrates the desired concept. In actuality, it is nearly impossible to do a perfect job.

An illustration of this practice is the after-Christmas sale. Many merchants use the two-for-one sale rather than a half-price sale. Their goal is to avoid storing the merchandise for 10 or 11 months. Not only does it take up space, it ties up unproductive cash. It's better to invest the money in high margin Easter merchandise than to let the seasonal merchandise sit in the warehouse.

When Will I Need to Borrow Capital to Facilitate my Cash Flow?

In our illustration of the cash flow circle, Figure 22–3, we indicated the first quadrant of the circle represented the purchase of materials for production and the second the wages and factory costs. When these expenses exceed the cash reserves of the business, the use of borrowed capital is necessary in order to carry the goods to completion, sale, and delivery. The sales will then return the business to the cash position and the ability to repay loans.

The management of production, scheduling, and sales activities makes the cash flow circle work. These cash flow projections, or plans, are necessary to negotiate operating capital loans.

How Does Credit Affect Cash Flow?

There are advantages and disadvantages in offering credit to your customers. It is up to you to weigh the pros and cons and to decide which will be best for your business. The use of credit affects your cash flow by delaying incoming cash.

Advantages of Credit:
- Customer loyalty
- Tendency of customers to purchase more
- Increased sales volume
- Competition with other businesses

Disadvantages:
- Higher cash requirements
- Need for meticulous records
- Expense of billing and accounting
- Nonpayment or partial payments

Credit Terms to Know

Open accounts—An open credit account allows a customer access to your inventory on credit terms with some kind of repayment agreement.

Installment plans—An installment plan is a purchase agreement whereby the customer contracts to pay a specified amount in increments, often monthly or yearly.

Interest charges—Interest charges are the cost to the customer for extending payment over a period of time. These are actually the seller's fee for allowing the buyer to use the money.

Interest free periods—This is the period of time after the buyer purchases a product or service with credit but before payment is begun on the *finance charge.* This means that the seller actually finances the purchase. Often the finance charge is hidden in the original cost of the product.

Revolving credit—This is a predetermined amount of credit, or credit ceiling, that a customer may use and for which he makes a minimum monthly payment on outstanding charges.

CREDIT REGULATIONS

The use of credit in business interactions is tightly regulated by the government. There are now Truth in Lending laws, which specify these regulations. In effect, they say that the business offering credit to a customer must openly and in writing state all credit arrangements having to do with the amount of the finance charges or interest rates. This is done by way of a *credit agreement* which the customer signs, indicating that he or she has read the terms of the finance agreement and agrees to abide by them. This legally protects the customer from unfair credit practices.

While Truth in Lending laws protect the customer, *credit bureaus* have been developed to protect businesses. Credit bureaus exist on both local and national levels. An organization called Associated Credit Bureaus, Inc. keeps information about nationwide credit transactions, which it makes available to businesses.

A business should have written credit policies which can be easily understood by both employees and customers. When a policy is written in simple but clear terms, misunderstandings are less likely to occur.

SCHEDULING PRODUCTION

How Will I Schedule Production Activities?

In any type of production business, your ultimate goal is to turn raw materials into products. To do this efficiently requires a lot of planning. It is one large process that is made up of many smaller components, or steps. When one of these steps is not operating properly, the other steps are delayed, and time and money are lost. These principles apply to plant and animal management as well as to manufacturing processes, Figure 22–4.

Once you thoroughly understand the process of production for your business, you will be able to operate at peak efficiency. Let's look at some of the studies that will help you understand your production process better.

The Production Layout.   How is your machinery organized? Where are things moved as the process goes from one step to the next? Is machinery laid out so that production is efficient or do your workers have to make many unnecessary movements? There may not be an ideal plan to organize your machinery if you manufacture more than one product. In that case, you need to look at several alternatives and select the most efficient one.

Factory Workload.   Will you be producing different items? If so, how flexible will your operation be? Will you be able to interrupt a long-term project in order to complete a short-term one?

Work and Time Standards.   Once your business is in operation, you will be able to study the effectiveness of alternate plans. You can compile data concerning time studies, performance records, and cost averages for any job done in a particular manner. With this information, it will be easier to project your capacity and costs for new jobs.

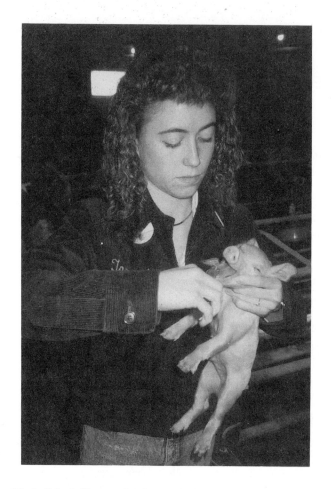

Figure 22–4   Scheduling production practices is important to successful swine management. (Courtesy David Buck)

Systems and Procedures.   With experience, you will be able to develop systems and procedures which will not only help your business run smoothly but will also save you money in the long run.

Variable Factors.   A business manager does not have full control of everything that affects business operation. Suppliers may not deliver on time. Customers may request a later delivery date. Shipping may be held up for a period of time. You can be prepared for such unforeseen delays by keeping a small inventory of surplus supplies, or by having warehouse space for storage.

Tools for Scheduling.   An operations process chart will outline the activities for any particular job. Before you begin production of a product or service, you should know, as exactly as possible, how much time and materials will be required, as well as what transportation costs will be.

Sequence Checklist

Scheduling

- A new job added to the schedule should only minimally disrupt currently scheduled jobs.
- The schedule should be flexible enough to allow for changes and unexpected delays.
- The schedule should be based, at best, on past proven data, and at least, on carefully made estimates.
- Extra flexibility should be built into scheduling if the production or service depends heavily upon outside suppliers or contractors.
- A backup system should be designed for the critical aspects of any operation.
- Scheduling should provide for maximum utilization of machines and skilled labor.
- Provisions should be made for overtime operations when schedules are interrupted or delayed for unforeseen reasons.
- The schedule should accommodate desired increases or decreases in output.
- Scheduling will affect quality of output.
- Scheduling plans should take into account the changes of fixed and variable costs at different levels of production or service.
- Even minor changes in scheduling can affect profits.

SOLVING PROBLEMS

How Can I Fix Production Mistakes?

Good management skills allow a person to plan operations in ways that tend to avoid problems. More important, a skillful manager knows how to identify problems that do exist (and they will, on a daily basis), and how to approach the problems in order to solve them quickly and effectively.

The *Small Business Administration* bulletin, ''Tips for Problem Solving,'' lists some potential problems that a business manager might expect to face. How would you handle the following situations?

- Something did not work properly and you don't know why.
- Something you need is unavailable, and you must find a substitute.
- Employees are undermining a new program.
- The market is not buying. What do you do to survive?
- Customers are complaining. How do you handle their complaints?

Problems are a part of life. Some will result from hasty decisions or improper planning. Others will be unavoidable, such as electrical outages or inflation rates. The success of your business will depend, not so much on the problems that come up, but on the way you solve them.

How Can I Learn Problem-Solving Skills?

There are some *logical steps* that you can take to approach problem solving.

Identify the Problem. This is not always easy. Ask yourself the following questions. What exactly was the anticipated outcome? What went wrong? Is the problem external or internal? Is it related to employees, equipment, markets, or operations?

Analyze the Problem. How significant is it? Does it have the potential of becoming a bigger problem? What could happen if you ignore the problem? Is it a problem that an employee could correct, or is it one that needs your direct attention? Is it a problem that has occurred in the past? Is it a problem that is likely to occur again? How have others resolved similar problems?

Find the Best Solution. List as many parts of the problem as you can identify. What steps need to be taken to correct each part of the problem? Compare the problem to others. What worked best in similar situations? Ask others who have faced similar situations. How effectively were they able to resolve their problems? What steps did they take to do so?

Take Action. Knowing the proper solution to a problem will not make it go away. Once you have designed a plan of action, you need to implement it. Who will you call upon to help you? Who will be affected by the problem or the solution? When will you take steps to solve the problem? Who needs to be aware of your plans? What equipment will be affected? Will you need to hire outside help, such as legal counsel?

Evaluate Your Solution. How well did it work? Would other ways have worked better? Faster? Cheaper? Would this solution work again in the future? What did you learn from the experience that will help you handle similar problems in the future?

The ability to solve problems is not inherent. It is a skill that we learn through watching others and observing the consequences of our own actions. Through the experience of making and correcting mistakes, we can become *adept* at solving problems, and, eventually, become better business managers.

SELF-EVALUATION

Scheduling Quiz

1. What do we mean by business scheduling?
2. Define and illustrate cash flow.
3. List the steps in problem solving.

STUDENT ACTIVITIES

1. Visit a local factory and observe how raw materials are handled and how inventory is shipped.
2. Observe work at a local garage. Notice how repairs are scheduled. Ask the service manager how many days ahead the work is scheduled.
3. Visit a local contractor and ask how work is scheduled when building several houses at the same time.

CHAPTER 23

How Can I Establish A Sound Set of Business Procedures?

OBJECTIVE

To explore good business procedures to be used by an entrepreneur.

COMPETENCIES TO BE DEVELOPED

After completing this chapter you will be able to:
1. Describe the importance of standard procedures and policies.
2. Identify the meaning of business ethics.
3. Identify appropriate items to include in a business policy manual.
4. Identify the minimum records a business must keep.

TERMS TO KNOW

Business policy manual

Delegation

Ethics

Fixed assets

Fringe benefit

Journal

Retrieve

Transactions

Communication

Depreciation

Fiscal year

Fortune 500

General ledger

Petty cash

Running balance

INTRODUCTION

Webster's Vest Pocket Dictionary defines procedures as "a series of steps in regular order." The procedures that you establish for your business will affect everyone you come in contact with—your customers, your employees, and your business associates. Before we begin to discuss the steps necessary to set up these business procedures, you need to think about your philosophy on doing business with people. In fact, that is your first, and most important, step in establishing a sound set of business procedures.

According to a December, 1986 story in the "Money" section of *USA Today*, "Interest in business *ethics* is booming. More companies are looking at the long-term value of doing business based on moral standards, not just on bottom-line reasoning." The story quotes a survey conducted by the nonprofit Ethics Research Center, Inc. in Washington, D.C., which found that 75% of America's 1,200 largest companies have established ethics codes. Another survey, done by the Center for Business Ethics at Bentley College, found that approximately 80% of *Fortune 500* service and industrial companies were working to institutionalize a business ethics plan for their companies.

The most important point that these surveys indicate to us as entrepreneurs is that there has been a change recently in the attitude that businesses are taking toward ethical procedures. You can benefit from the lesson that many large companies have had to learn the hard way. In years past, a prevailing attitude among many companies has been, "Nice guys finish last." In light of several business scandals that have caught public attention recently, companies are rediscovering that ethical policies are good business. These policies can be as important to a business's success as the financial management plans that must be implemented.

What are ethical business procedures? We could define ethics as the rules or standards that govern the values and conduct of a group of people. Ethics covers a large scope of behavior. In business, bad ethics could refer to grand-scale fraud or the practice of pocketing company pencils. You must provide the guidance for your business's policies. Policy for company procedures needs to stress records and how to handle them. This includes the records that are essential and how they are recorded and preserved.

CASE STUDY

The Case of the Greasy Palm

Craig Donovan has been running his own office supply business for 16 months. His major concern during the first year had been to develop his customer base, so he focused his attention on advertising and sales. He was pleased with the way the business was developing. Because of Craig's inexperience in retailing, he had hired employees with prior experience in their respective departments. Craig felt especially fortunate to have found Alex Johnson, who had handled purchasing for a comparable business for several years. Alex seemed to know all the ropes. He even knew the sales representatives from many of the suppliers that Craig wanted to use. Because the store sold a wide variety of products from different suppliers, purchasing wholesale items was not only complicated but highly competitive. Craig was glad to have that aspect of the business under Alex's experienced management so he could give his attention to other important matters.

One day Craig received a call from a salesman for a new supplier. As usual, Craig politely suggested that the salesman speak to Mr. Johnson, head of the purchasing department. "Mr. Donovan," the salesman replied, "I have spoken with him. I'm new at this, and I work on a commission. I believe I have a product that you would be interested in, but I can't afford to get inside your door."

Craig was intrigued and a bit confused. He invited the salesman to come in and speak with him personally. Craig and the salesman talked in the privacy of Craig's office, and, to his dismay, he discovered that the salespeople for the companies that supplied his business regularly gave "incentive favors" to Alex Johnson. "It's not exactly standard practice," the salesman explained, "but it happens. And Mr. Johnson has learned over the years who is willing to play the game. Ask him where he got his new microwave oven, or if he recently received a clock radio. I have a friend working for a competitor. He told me that in order to sell my products to your business, I would have to first grease Alex Johnson's palm. I can't afford to do that, and I wouldn't even if I could."

Craig was shocked. A confrontation with Johnson proved the salesman right. Alex Johnson was terminated as an employee of Craig's business.

What Does the Case Study Tell Us?

- Craig realized that as owner/manager of his business, he is responsible for all aspects of the business.
- Craig also learned that he shouldn't assume that everyone shares his attitude toward integrity. Different people have different values. It was up to him to inform his employees of the standards that he expected from his employees.

ESTABLISHING BUSINESS STANDARDS

Too often, standards of conduct in a business are not spelled out for employees. New employees have to work awhile to get "a feel" for the company's values. Sometimes, as in the case of Alex Johnson, they never do. *Communication* among you and the people that you work with is imperative.

The Business Policy Manual

How can you communicate your expectations to your employees? One way is to develop a *business policy manual*. This has several advantages.

First, it helps you to clarify your attitude toward ethical business procedures. You need to think about how you want to structure your relationships. How will you best meet your customers' needs? Your employees' needs? Your banker's or suppliers' needs?

A second advantage is that your employees will know exactly what is expected of them. What is your attitude toward personal telephone calls for employees? Or toward the use of office equipment (typewriters, copy machines, and so on) for personal use? You may want to offer this type of thing as a *fringe benefit* of the job, or you may decide that such practices become too expensive and should be avoided. The only way your employees will know is if you communicate your policies to them. The time to do that is when you hire them. You will find it easier to begin with a clear policy than to change things after undesirable habits have developed.

A third advantage is that you will have a sturdy base on which to build trustworthy relationships with business associates. They will know they can expect consistent responses from any of your employees. It's frustrating for anyone dealing with a business to feel that they will get varying answers depending on which employee they happen to speak with. Your employees represent you, and they can do that best if they know what your attitudes are. High standards of work are essential to business success. Tammy follows high standards, Figure 23–1.

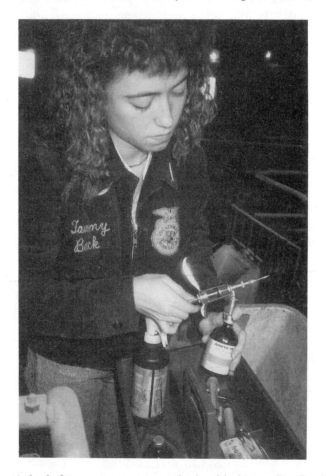

Figure 23–1   The standard of top management practice is critical to profit. (Courtesy David Buck)

Perhaps the greatest advantage will be in your relationship with customers. When establishing your policies, consider their needs. They are paying *all* the bills. Regular customers aren't attracted by price alone. They want to deal with a company that is consistent, reliable, and trustworthy. A sound, well-understood policy manual can assure that they receive that kind of treatment.

A policy manual can be as simple as a few typewritten pages copied for each employee. It can also be an elaborately printed, illustrated brochure. Remember that the packaging is not as important as the content. You might begin with a broad statement about your company's goals. In what direction are you headed? What do you foresee in the future—five or ten years from now? What is the company's growth potential? What sort of image do you want to project for your customers and others?

At this point, details become important. Not only will employees want to know your policies, but they have a right to know what is expected of them. The following is a list of items that you might consider including.

1. *Working Hours.* What are the regular working hours? Will you be flexible or do you expect your employees to clock in and out at precise times?

2. *Overtime Hours.* Will employees be asked to work overtime? How will they be compensated? Will you understand if they are unable to comply?

3. *Salary Schedule.* When, and how often, will employees be paid? What deductions can they expect? Will you pay them wages or will you offer them a salary?

4. *Benefits.* Employees are interested in the benefits your company will offer. They will want to know about:

Vacations	Retirement plans
Sick leave	Disciplinary policies
Leave of absence	Employment termination
Holidays (paid or unpaid)	Resignation
Insurance benefits	Discharge (and reasons)
Worker's compensation	Layoffs
Medical insurance	
Life insurance	

5. *Customer Relations.* How will employees approach customers or answer the telephone?

Many of these items have been discussed in detail elsewhere in this text. They are some of the topics you should address in your policy manual.

Training Programs

New personnel go through a training period. This may be informal—a time during which they get to know the other employees and become familiar with the operating procedures. On the other hand, the position may call for a structured and intensive training program. Before you hire a person to fill a position, you need to know exactly what they will be doing and see that they learn to do it. The ideal situation would be for you to lead them step-by-step through all of their duties, but that is neither realistic nor practical. Therefore, you need to develop a training program for new employees, either using experienced personnel or written instructions.

Hiring Procedures

From the Beginning.   Once you have thought seriously about the standards of conduct that you expect from your employees, and you know how to communicate these standards to them, you are ready to begin hiring personnel. Interviewers will tell you that it is not safe to rely solely on your memory when the time comes to make final hiring decisions. You want similar information from each applicant so that you can make fair and reliable comparisons. The easiest way to do this is to develop an application form that will give you the information you need about each person. A sample form is given in Figure 23–2. This form is very general. Each business has different needs—there may be information that you will need from applicants that other businesses are not interested in having. At the same time, you may not care about other qualifications. You will want your application form to reflect your business. Don't ask a prospective employee if he or she is bondable if you don't ever anticipate needing a bonded employee. On the other hand, if you want your company to be active in community affairs, such as charity drives, you may be quite interested in their past community activities.

Once you have several applications, you can select the people that you would most like to hire. You may need to spend a day or two interviewing applicants. This process can be very unsettling for those being interviewed. It may not be easy for you, the interviewer, either— especially if you are not well-prepared for the interview. You will want to know in advance what sort of questions you wish to ask. You might want to take brief notes during the interview. Give the interviewee an opportunity to relate any special skills or to make comments. After the interviewee has left your office, make notes of your personal impressions.

Legal Compliance.   There are laws that govern hiring practices. There are questions that you, as an employer, are not allowed to ask, either on your application form or in an interview. Because these laws change from time to time, you need to consult your lawyer before you inadvertently ask illegal questions. Currently, there are federal anti-discrimination laws which prevent discrimination in hiring.

The Equal Pay Act of 1986 mandates equal pay for women. Title VII of the Civil Rights Act of 1964 prohibits discrimination on the basis of race, sex, color, religion, or national origin for employers with 15 or more employees during 20 weeks of the calendar year.

The Age Discrimination in Employment Act of 1967 prohibits hiring or firing because of age for persons between the ages of 40 and 70. This applies to employers with 20 or more employees for at least 20 weeks of the calendar year.

Executive Order 11246 requires no discrimination in employment practices on the basis of race, sex, color, religion, or national origin for employers with federal contracts or subcontracts of $10,000 or more.

As you can see, the laws are complicated. It's important that you are aware of them. Although your lawyer's consultation fee may seem like an unnecessary expense, it is economical compared to the cost of legal action taken against you.

In-House Communications

You may have so many employees that you will be unable to talk with each of them frequently. Although it is good to keep in touch with your employees (you want them to know

EMPLOYMENT APPLICATION

PERSONAL

Name: _____   Social Security Number: _____

Address: _____   Phone Number: _____

_____   Date of Birth: _____

EDUCATIONAL BACKGROUND

Schools Attended	Dates	Degree Awarded
1.		
2.		
3.		

Other Training (Technical, Trade, Business, Military):

EMPLOYMENT HISTORY (beginning with most recent employment)

Name/Address	Dates	Salary	Duties	Reason for Leaving
1.				
2.				
3.				
4.				

REFERENCES

Name 3 persons we may contact who are familiar with your qualifications:

Name	Address/Phone #	Relationship
1. _____	_____	_____
	_____	_____
2. _____	_____	_____
	_____	_____
3. _____	_____	_____

ADDITIONAL COMMENTS (Include any special qualifications, skills, etc. Use the back of this page if necessary.)

_____   _____
Date   Applicant's Signature

Figure 23–2 A sample employment application. (Courtesy The Three Entrepreneurs)

that you are interested in their work), other demands on your time may preclude frequent discussions. One method of handling in-house communications is the memo. A memo is a quick and efficient way of getting information to everyone. An advantage to this is that it provides a written record of communication. Word-of-mouth messages that get passed from employee to employee can become altered. A written communication forwards your message in unquestionable terms to everyone concerned. In addition, encourage your employees to communicate with you through written messages. Although you don't want to do away with personal contact, memos can save both you and your employees many steps and valuable time.

Record Keeping

If you have had accounting or bookkeeping courses, you will have a good idea about the kinds of records you want to keep as well as the type of bookkeeping system that you want to use. In this text we seldom give you hard and fast rules about running your business. Our tips about record keeping are suggestions that will help you get organized. As in other aspects of your business, you will need to adapt any system to your needs. Tammy Buck keeps litter records, Figure 23-3.

The more simply you design your record keeping system, the more easily you will be able to *retrieve* important data or documents when you need them. It's important to be thorough and accurate, but avoid unnecessary complication.

Your system should be easy to understand. Remember that you are not the only person who will be using it. Others will be filing records and recording data. As long as they understand and use your system correctly, you will have no problem finding information that someone else has stored. A well-designed plan for record keeping will help everyone involved to be consistent.

You want your records to be easily accessible. File cabinets are a necessity today. As tax laws and business ventures become more complicated, computerized record keeping is becoming the easiest and most convenient method of keeping reliable and consistent records. Avoid putting your records in boxes that are stored in hard to reach places. It may be nice to have mountains of paperwork stored out of sight, but you pay a price in valuable time when you need to retrieve information that is not handy.

Minimum Record Keeping.   There are some records that you must keep in order to run a successful business. These are:

1. *Sales Records.* Keep a record of all sales (retail and service) that your company makes. This information should include the party to whom the sale was made, the date, the amount and the item of sale.

2. *Cash Receipts.* Your sales will not be exactly the same as your cash receipts. We discuss this in more depth when we talk about cash flow. If you have sold an item on a time installment plan, or for later payment, then you need to keep track of the payments that are made for that item.

3. *Cash Disbursements.* You need to have a record of all cash paid out by your company. This should include the date, the amount, the receiving party, the reason, and the check number (as well as the account if you have more than one account from which you pay bills).

Figure 23–3 Tammy's sister watches her notch the ear of a baby pig to facilitate record keeping. (Courtesy David Buck)

4. *Accounts Receivable.* This is a record of all accounts that are not paid in full to you. This needs to be kept according to the due dates of the payments. It makes sense to keep a record of all current accounts due, as well as accounts due in 30, 60, and 90 days. A reliable accounts receivable record will be invaluable to you as you plan the cash flow for your company. (It will also probably be a requirement from your lender.)

5. *Sales Tax.* You need to know exactly how much sales tax you have charged your customers as well as how much sales tax you have paid on items your company has purchased.

Balancing the Books.   Maintain a *running balance* for each of the different types of records that you keep. In other words, you'll need to have daily, monthly, and annual totals for each set of records.

You will also need to balance your books with your bank statements monthly. Mistakes do happen, and are easier to remedy if caught early. When your books don't balance with your bank statements, either your company or the bank is in error. Either way, you need to know as quickly as possible. Mistakes can be disastrous, especially if you are dealing with large amounts of money or are operating on a very tight cash flow.

Petty Cash Funds. Depending on the types of *transactions* that you make, you may find a *petty cash* fund convenient. If you make several small purchases every day, it will become expensive and time consuming to write a check for each. Some businesses store small amounts of money in a safe place in the office, and use it to pay for small items. If you do that, keep a running record of the amount deposited in petty cash and of the amounts withdrawn. At the end of a certain period of time, perhaps when you deposit additional cash in the petty cash fund, you can make a summary of the small transactions that came from there.

Company Purchases. It is also a good idea to keep a file of all the equipment purchased by your company. If you list the equipment, the date it was purchased, the name of the supplier, the amount of the check and the check number, as well as any maintenance contract that was purchased at that time, you will always have a record for insurance, replacements, and so on.

Insurance Records. You will want to keep a record of all the types of insurance policies that your company has purchased. Include the type of coverage, the name of the insurer, the date purchased, the date of expiration, and the annual premium.

Depreciation. You should keep a *depreciation* schedule for all depreciable assets of your business. Tax laws change, so consult with an accountant to ensure accurate records. Straight line depreciation is figured by taking the value of the item and dividing it by the expected life of the item. For example, if you purchase a computer program for $600 and you expect to use it for 20 years, its value will depreciate $30 a year. At the end of six years, the value that you assign to your software will be $420. Actually, given the rapid changes in computer software, it's not likely that you'll find a program that you will want to use 10 or 12 years from now. However, the example does illustrate the ease of figuring straight line depreciation schedules.

There are other types of depreciation schedules. Some deduct greater amounts for the earlier years of the life of the equipment. Again, your accountant can help you figure which depreciation schedule will best serve your needs.

Financial Statements. You will need a system to record current financial statements. You will also need to have access to balance sheets for any given time. These include a list of current assets and liabilities. An up-to-date income statement which lists all profits and/or losses for any given time period is essential.

Payroll Records. You will not only need this information for your cash flow planning, you will also need it when you file your tax statements. You will probably use this information to make both quarterly and annual state and federal tax reports. At the end of the year, you must distribute W-2 forms to your employees, which will show their total pay and the amount of tax dollars withheld from their paychecks.

Employee Information. Many employers keep a file of information on each of their employees. They keep records such as address, phone number, social security number, next of kin, rate of pay, number and type of exemptions, and hours worked.

Records for Taxes. There are not many adults in our country who don't know the significance of April 15. It is the annual deadline for filing income tax records. (Note that a business files tax returns on the annual basis selected at the start of the business.) For the organized business person who has kept thorough and ordered records, this date is only another deadline. To the business person who has kept haphazard records throughout the year, this date means hours of scrambling to gather and retrieve important records.

Business managers develop their own system of keeping tax records according to their needs. You need to be able to retrieve all data pertaining to the cash flow of your business—what money was taken in, what money was paid out, and for what reasons. You also need to be able to substantiate those claims if the Internal Revenue Service asks you to do so. This means that you need to have two sets of records: one in which you keep track of cash flow (including the reasons), and one (such as a filing system) in which you store documents that offer proof of your claims. The following are some of the places that this type of information is stored:

1. *Checkbooks.* It is wise to run all funds, both incoming and outgoing, through your checking accounts. This becomes a reference for what moneys were paid out or deposited, the dates, and the reasons. A canceled check offers proof of payment in the event of an IRS audit. It is important, therefore, to keep your personal financial transactions separate from your business transactions. Even if you are running a small business out of your home, keep separate checking and savings accounts for your business expenses and income.

2. *Cash Receipts Journal.* This is no more than a quick and handy list of money that you have received, the date, the party, and the reason.

3. *Cash Disbursement Records.* This is another ready reference summary of all money paid out, the date, the party, and the reason.

4. *Fixed Asset Records.* A list of all *fixed assets* (your buildings, vehicles, equipment, and so on), the date of purchase, the amount paid, and the current market value.

5. *Wages.* A record for each employee listing the date and amount of all wages paid, the amounts withheld for taxes, the deductions for benefits (such as insurance and retirement funds).

Corporations and Partnerships

Records required for corporations and partnerships are more extensive. The manager of a corporation must keep records of:

- Salaries paid to officers and
- Dividends paid to stockholders.

Each partner in a partnership must keep records of their individual business profits and losses.

Permanent Records

Some records need to be kept for a period of years. Your accountant can advise you about the length of time required for each type of record. The following is a list of documents that you should plan to keep accessible for several years:

- Cash books
- Depreciation schedules

- *General ledger*
- *Journals*
- Financial statements
- Audit reports
- Copies of tax returns

At the end of each *fiscal year,* these records can be compiled and placed in permanent (but accessible) storage.

Accounting Services

If your business is very small, and you have the time and knowledge to manage your own accounts, you may not need the services of an accountant. However, many business managers find these services indispensable.

There are two options for the use of an accountant. Many large businesses hire their own in-house accountant. This is a salaried employee (or employees) who works full-time keeping business records, doing payroll, and preparing tax returns. Smaller businesses that don't need the services of an accountant continually often hire an outside firm to prepare their tax statement. An outside accounting firm can also give you advice on setting up your record keeping system. They can tell you what vital information must be recorded, prepare your financial statements, and help you make decisions about major purchases or expansion. Remember, attention to little things adds up, Figure 23–4!

DELEGATING WORK AND RESPONSIBILITY

From the beginning, this enterprise has been your ballgame. Your ideas have created your business and you have made all the decisions. What happens when you no longer have the time to control all the details involved in running your business? Who will "mind the store" when you are out of town? The time will come when you will need to share the responsibilities of managing your business. A manager does not do all of the work. Rather, the manager oversees the operations that get the work done. He or she does this by delegating the work and responsibility.

It is hard for an entrepreneur to let go of the control maintained since the inception of the business. The key to delegation is the personnel that you hire. When you hire personnel to fill your management positions, ask yourself these questions: Does this applicant have the ability to manage people and operations? Can he or she take the initiative to get things done without prompting from you? Can the prospective manager motivate others? When you have hired personnel who are competent and conscientious, you will be more willing and better prepared to "turn loose the reins."

The next step in *delegation* is to specify responsibilities. What decisions will be made at what level? Who will eventually hire or terminate employees? Who will determine new purchases? Who will authorize vacations? Who will be responsible for the many small decisions that must be made daily?

Figure 23–4   Attention to details adds up to profit. (Courtesy David Buck)

In order to answer these questions, you must coordinate the departments within your business. The decisions that you make while your business is young will create the base for procedures as your business grows. If you want your business to develop, you need to prepare for the time that it will become departmentalized. For instance, you may eventually have a production department, a sales department, a purchasing department, and an accounting department, as well as an administrative staff.

Management and supervision operations within a business often work on a set of many levels. Think of these levels as a triangular system in which the highest level positions control the most comprehensive decisions, while the lower levels control the smaller decisions. The employees at the lower level of the triangle are your work force. They are the employees that contribute the labor to produce your company's product or service. Figure 23–5 is a simplified illustration of one organizational option.

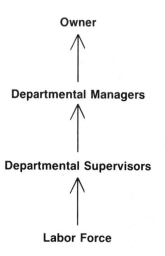

Figure 23-5 Levels of organization. (Courtesy The Three Entrepreneurs)

Maintaining Control

The advantage of delegating responsibility is that you free yourself of the time required to oversee all daily operations. Many decisions can be made by others, but you are ultimately responsible for all aspects of your business. You want to know what is happening within your company—what decisions have been made, and why.

A procedure that allows you to keep on top of things is the practice of periodic reporting. Weekly or monthly staff meetings in which activities and proposals are discussed keep everyone informed about current and future business plans. Staff meetings also offer an opportunity for problem-solving sessions.

The key to good management practices is *communication*. Encourage your staff to communicate with you and with each other. When meetings are inconvenient, encourage your staff members to communicate with you through memos. The important thing is that everyone must feel free to discuss important issues with the appropriate person when necessary. It is important that you emphasize to your staff the value of listening. Encourage them to be available to all employees for open communication.

As you delegate responsibilities, remember to give your staff the opportunity to use their best judgment. Allow them to be creative. Remember to judge their efforts by the results that they produce rather than the methods that they use, provided the methods used do not conflict with the company's code of ethics.

SUMMARY

The primary goal of good business procedures is to help you manage the endless small details involved in business management. Failure to pay attention to even small details can erode profitable operations. When the details flow smoothly, larger operations tend to work more efficiently and your job as owner/creator of a business will be easier as well as more successful.

SELF-EVALUATION

Business Procedures Quiz

True-False: On a separate sheet, write T for true statements and F for false statements.

1. Business ethics refers to how a business pays its bills.
2. You could define ethics as the standards that businesses hold to.
3. Payoffs and kickbacks are legal but morally wrong.
4. Policy refers to written or unwritten statements that define the type of actions or standards to be followed by the business.
5. Policy statements are useful to communicate your standard of ethics to employees and others.
6. Writing a policy manual will not help you set your standards.
7. Such a manual will provide consistency between people in the organization.
8. Working hours will not appear in the policy statement.
9. Policy on leave and vacations will be an appropriate part of the manual.
10. The business policy manual should include the company's philosophy and goals (immediate, short-, and long-term).
11. Legal compliance statements are not appropriate in the policy statement.
12. Communication is more important in a large business than in a small business.
13. Communication is easier in a small firm.
14. A minimum records system must have enough information to satisfy the Internal Revenue Service and the business owners.
15. Petty cash funds are not legal.

STUDENT ACTIVITIES

1. For your proposed business, design:
   - a letterhead
   - a sales slip
   - a monthly billing statement
   - a company memo—message and reply form (3 copies)

   If computers are available, use their capabilities.
2. Practice completing simple sales slips and other common business forms.
3. Use the tax form for depreciation and complete a depreciation schedule for your proposed business.
4. Obtain and complete an employment application.

CHAPTER 24

How Do I Develop Customers?

OBJECTIVE

To identify a customer profile and customer development strategies.

COMPETENCIES TO BE DEVELOPED

After completing this chapter you will be able to:
1. Identify three ways to recognize desirable customers for a business.
2. Develop a profile of desirable customers given the type of business or service performed.
3. Explain the concept of public relations as compared to publicity.
4. Suggest ways of involving customers in promotional programs.

TERMS TO KNOW

Bonded
Emphatic
Patrons
Public Relations (P.R.)
Quality time
Runts
Targeted prospects
Unflagging

Built-in bias
Human relations
Potential
Prospective
Research farm
Strategies
Word-of-mouth

INTRODUCTION

Every business depends on its customers whether it is a service, sales, or production type of business. Developing and maintaining customers is an all important consideration. In this chapter we are going to examine customer-related activities as contrasted to the promotional activities described in another chapter.

CASE STUDIES

The Case of Proof

For several months, Phil had been trying unsuccessfully to sell Walter his line of feed. Phil had had two of Walter's sons in class previous to starting the Exeter Feed and Grain business. Phil had thought that he and Walter had a mutually trusting relationship but suddenly Walter appeared to be suspicious of Phil's motives. To overcome this resistance, Phil suggested that they conduct a head-to-head feeding trial with two lots of barrows. One would be fed the current brand of supplement and the other, Phil's brand. The sons were enthusiastic so Walter agreed. Two corn bins were leveled out and measured, and one was used for each of the pens. When the lot fed Phil's brand went to market first with less feed consumed, Walter was sold and soon was telling his neighbors about the good record on the new brand. This successful trial on one farm led to other trials and Phil was able to increase his customers on this basis. Such *word-of-mouth* support for good service is good for business. Ronnie, Figure 24–1, gives good service.

Figure 24–1 Ronnie and Jay inspect the work in progress on Jay's truck. Ronnie was awarded the State FFA Degree for his business records. (Courtesy Ronnie Orem)

The Case of Extra Service

Don runs a centrally located tire sales and service business with a fleet of trucks run by well-trained, courteous operators. Each time one of Don's employees services a vehicle, a simple safety check is performed to alert the owner to needed service. Since Don has no vested interest in such repairs, the response is positive and customers return or are *bonded* to the tire company.

What the Cases Tell Us

- People often distrust businesses.
- Visible results of the product's worth are effective sales tools.
- Extra services that have no *built-in bias* are good *public relations*.
- Word-of-mouth advertising is good when it is positive.
- Customer development takes *quality time.*
- Features that cause your business to stand out from the competition will build customers.

DEVELOPING CUSTOMERS

Going After Desired Customers

Early in Phil's career in the feed and grain business, his block man (or company sales representative) conducted a meeting for local dealers. In the course of the meeting, the dealers were told how to go after the key livestock feeders in their areas. Phil and his partners decided to try the suggestions so they sat down and made a list of *potential* customers based on the following criteria:

- Size of operation,
- Standing in the community,
- Credit rating or ability to pay in a timely manner,
- Their knowledge of the feeder,
- The quality of management performed by the feeder.

Armed with this list, they each thought of ways to make favorable contacts with the feeders. It was obvious to them that time spent selling to a feeder who used 60 tons of feed would be more profitable than the same amount of time devoted to a feeder who only needed 6 tons of the same feed. Also, adding feeders with poor payment records or high debt levels would not be beneficial to this young business.

Techniques of developing loyal customers will vary with the type of business. However, Phil's procedures can be applied to many businesses and can serve here as a working example. Many of the techniques described will work with a wide variety of enterprises.

The key feeder list was shared with the company salesman and he and Phil visited the top three prospects on the list. Later the salesman returned to call on them when he was in the community. The feed company operated a *research farm* at some distance from Exeter. An expense paid trip was arranged for these *prospective* customers so that they could see firsthand the research being conducted. When the county fair was held, the son of one of these families showed the second place steer, which Exeter Feed and Grain purchased at the auction which followed. The father of this boy was later invited to attend a state university beef feeding seminar with Phil and two other customers. It is not surprising that 60 tons of bulk cattle feed were delivered to this family's farm the following year. Similar *strategies* were utilized for each of the prospects on the list. Other customers were not neglected. Instead, many efforts were made to improve their management practices so their profit margins would encourage expansion. Personal service, including after hours contacts, are powerful helps. Figure 24–2 illustrates the help of family members.

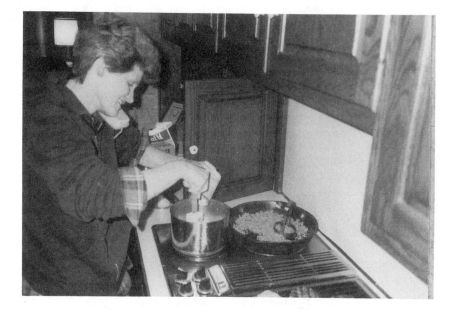

Figure 24–2 Family members are important to an entrepreneur. Janet handles a customer's call at the customer's convenience. (Courtesy Janet B. Connelly)

You will want to make a plan to identify the type of customer you want to attract and develop. Identify the key characteristics of the desired customer and then identify specific customers and a means of selling your product to them. The plan may include ways to get customers who usually buy only one item to buy other items you stock. Set a timetable for your plan and then set to work. Remember the lessons learned in the goal-setting chapter and make your goals challenging but obtainable!

Some elements to include in your development plan include attention to public relations. Specifically include:

1. Getting better acquainted with present customers,
2. Getting to know what these customers want to buy,
3. Maintaining highest quality service,
4. Calling on the targeted prospects,
5. Involving the prospects in some promotion, and
6. Using customers in your advertising program.

Getting to Know Your Patrons.   Learning to call your customers by name will pay handsome dividends to most businesses. In addition, it is important to know something about their use of your service or product. You may have an opportunity to suggest better ways of utilizing the item currently purchased or even items you sell that the consumer is not now using. An understanding of the customer will aid in these sales efforts.

Human relations are important to goodwill and sales. Taking a sincere interest in your customers is time well spent. When you know them well enough to inquire about their health, family, and activities, it will generate good relationships that often translate into regular patronage or additional sales. We tend to like those who show an interest in us.

When the first cars were delivered to a certain dealer after a long strike, he took advantage of the demand and sold at very high prices. One of his friends told him that if he continued trading in this fashion, he soon wouldn't have any friends left. His reply was that the way things were going, he wouldn't need any friends. He was a misguided individual. We are primarily social beings and as such we need friends. The need varies from person to person, but for the most part we need other people and social approval.

In addition to recognizing the customers at your business, you need to greet them in other settings. Greeting your *patrons* by name at other times and places is a powerful public relations tool.

Public relations, when defined as "living right and getting credit for it," is important to everyone. Sincerity is important in all human relations; rarely can it be successfully faked.

Maintaining the Highest Quality Service.   Reliability and dependability are important business virtues. If your service is contracted for a certain time, that time should be honored. If a customer has livestock and the feed is needed that day, a delay may cause the customer to lose production. Phil would make deliveries after the business closed if they couldn't be completed before five o'clock. It is important to live up to business commitments.

Using Customers in Advertising.   Phil used pictures in his advertising whenever possible. He was careful to obtain permission for any customer's picture he used. These ads were popular because they featured friends and neighbors. Customer proof was used with the data featuring the trials at the customer's farm. Almost everyone enjoys publicity in the local paper, as long as it is positive. Figure 24–3 illustrates the idea.

Calling on Targeted Prospects.   Visiting high potential prospects should be a priority. Some of the methods Phil used to develop contacts have already been discussed. Note that contacts can be developed on the basis of mutual organizations or activities. In addition, mailings can initiate contacts. Remember the young entrepreneur who learned too late that the sign he purchased could not be installed? He thought of a way to bring new customers to his restaurant. He mailed the *targeted prospects* a coupon good for a half-priced second entree good for a specified time. He then stamped it and returned it to the customer so that it could be used a second and third time during the promotion. This created a lot of goodwill.

Involving the Customer in Some Promotions.   Phil regularly used in-store promotions with two live animals or a dozen layers in cages. In one of the promotions using two pigs, weekly records were displayed above the pens. On Saturday, customers were asked to witness the weighing for the records. The pigs were weighed a dozen or more times each Saturday before the weights were officially recorded. This technique involved a number of people in a more

Figure 24–3 Using customers in your advertising creates good will. (Courtesy Lester Bean)

emphatic way than the record alone would have. It caused these people to feel involved and they would drop by another week to see how the trial was going.

Another technique used by Phil was to purchase the animals he wanted to use in the store from a prospective customer. Pullets put into the laying cages would have their production records displayed and that prospect could compare the records to those of his remaining pullets. For a pig demonstration, Phil purchased the *runts* and practiced the sanitation and fortified feed methods needed to turn them into profitable animals. These demonstrations helped develop customers. The source of animals used in demonstrations was never revealed unless the owner agreed. After the animals were thrifty and showing excellent results, such permission was frequently given.

Getting to Know What Your Customers Want to Buy.   Another useful tool in developing customers and building business volume is getting to know what other items customers would like to buy. Early in Phil's tenure at Exeter Feed and Grain, a customer who had purchased some chicken feeders asked Phil if he had any hog waterers. Phil asked him what type he would like and offered to order them at a 10% discount since he did not have them in stock. Sales were increased and a satisfied customer served.

Phil decided that a brief questionnaire would help him discover what lines, in addition to the feed and sanitation products, the infant business should add to the existing elevator grain business. From this survey came the mineral and salt blocks, molasses feed, large hog and cattle feeders, and other items that were then stocked. Another way of knowing what your customers would like to purchase is to listen to what they ask for while in the store. The items your competition stocks may be clues to possible profit producing inventory.

SUMMARY

Developing customer loyalty is a challenge to every business and it is best met with honesty, candor, and *unflagging* consideration for the customer's best interest. In the long run, this will be in the business's best interest. A good sale is one that is good for the customer and the seller. It helps the customer to obtain the greatest advantage from the service or product. When your customers are convinced that you are sincere in wanting them to get the greatest advantage from their purchase, they will respond positively. In fact, they will likely tell their friends about your business.

This word-of-mouth advertising will go far in building your pool of customers. Everyone expects you to say your service is the best, but when someone else says it, it carries more weight.

In customer development, personal contacts are key elements. Although it is impossible for the "boss" to handle every customer, each must feel important to the business. These contacts must be maximized at all times.

SELF-EVALUATION

How Do I Develop Customers?

True-False: On a separate sheet, write T for true statements and F for false statements.

1. Many people distrust business persons, often unjustly.
2. Word-of-mouth advertising is always good advertising.
3. Time is a vital ingredient in customer cultivation.
4. A business person should spend as much time with a customer who will buy two items as with one who will buy 20.
5. A business owner should get acquainted with the customers.
6. Business owners should cultivate a good memory.
7. Reliability is an important business asset.
8. Customers should not be used in advertising.
9. Honesty is not always the best policy in business.
10. There is no ready substitute for personal contacts in developing customers.

STUDENT ACTIVITIES

1. Look at how manufacturing businesses in the area develop customers, or how retail businesses promote customer development. Use care in pursuing your studies. You will not want to approach local businessmen and women with poorly thought out questions.

2. With your classmates role-play a businessman asking a customer to be in an advertisement.

3. Collect newspaper clippings of advertisements and determine their method of attracting customers. These methods include quality, price, service, gimmicks, and so on.

CHAPTER 25

How Do I Manage Personal Contacts?

OBJECTIVE

To identify ways of improving business contacts.

COMPETENCIES TO BE DEVELOPED

After completing this chapter you will be able to:

1. Discuss the importance of human relations.
2. Identify the human relations skill that can be used by everyone.
3. List three ways an individual can improve his human relations skills.
4. List multiple ways a business can improve its human relations skills.

TERMS TO KNOW

Break	Cloudburst
Demonstrations	Feed efficiency
Happenstance	Human relations
Image	Sanitation products

INTRODUCTION

Public relations was defined by a noted speaker as "living right and getting credit for it!" It is important to differentiate between publicity, an effort to call attention to one's enterprise, and public relations, the development of public awareness through favorable means and the development of favorable responses to the enterprise.

Public relations, therefore, deals with fostering goodwill with the public—the community, employees, customers, and even the competition. These activities are conducted as a planned activity and not as *happenstance*.

CASE STUDY

The Case of a Flooded Business

Phil's business was located equidistantly between Glass River and Grass Creek. During late July, a *cloudburst* hit the Grass Creek watershed when the Glass was already at high flood stage. As a result, Phil's business was flooded to a depth of 6 feet for an hour or two.

The company that supplied the feed and sanitation products came to the aid of Exeter Feed and Grain. They immediately had Phil take all his inventory to the local landfill, and they alerted the local newspaper which pictured the dumping. They then replaced the total inventory and took out a full page ad for three weeks telling customers that they would get undamaged feed and *sanitation products* at Exeter Grain courtesy of the XXXX Company because they valued their customers.

What the Case Tells Us

- Public relations is a planned way of maintaining public approval.
- The XXXX Company not only developed positive feelings with the Exeter community, it developed strong loyalty from Exeter Feed and Grain by being a good neighbor.
- The "good neighbor" deeds of the supplier built goodwill for later business that maintained or even improved the share of the market.
- Publicity can be obtained by contacts that are more powerful than paid advertising. The news story ran on the front page with a picture clearly showing the company's name on the truck.

MANAGING BUSINESS CONTACTS

Developing a Business Image

Developing a business *image* is the beginning of good public relations. How did Phil manage his business contacts to develop a desirable image? Here are some of the principles Phil maintained from the day the business started:

- The customer was given the *break* of the scale.
- The customer was invited to make the weight ticket.
- Phil learned the operations of his customers—he visited their home farms.
- Phil kept his agreements—he delivered on the date promised even at the expense of delivering after the store closed.
- Phil promoted business by having expense paid tours of the XXXX Company's research farm.
- Phil helped farmers conduct *feed efficiency* trials.
- Phil conducted in-store *demonstrations*.
- Phil featured customers whenever possible in his advertising.
- Phil took farmers to a nearby restaurant for coffee whenever he could.
- Phil used the company symbols on the buildings and trucks.

- Phil used the company symbols on license plate tags and gave them to customers.
- Phil was asked to run for city council and agreed to serve. He was elected.
- Phil was active in a local church.
- Phil was an active member of the local service club.

Promoting Good Human or Public Relations

There are a number of good practices to follow to improve your *human relations* skills. They include but are not limited to:

- Smiling
- Greeting
- Listening
- Helping
- Performing
- Serving

Smiling. Smiling is always appropriate when it is sincere. Someone has noted that it takes fewer muscles to smile than to frown. A warm smile is worth cultivating. When you smile, others smile back and it is a positive feeling. Your enterprise will run more smoothly when you smile as you work. Smiles are positive reinforcements to those around you and a self-starter for you. You feel good when you smile. Enthusiasm is a positive asset, Figure 25–1.

Figure 25–1 How you work with people is important. Bob is enthusiastic as he describes his system for Western students. (Courtesy Robert L. Maudlin)

Greeting. Greeting people can be a business asset and doubly so when you greet them by name. Learning customers' names will help bond customers to your business. Learning of their special interests, their work, and family is time well spent.

Listening.   Listening is a valuable skill poorly practiced by most people. How well do you listen? Five minutes after an introduction can you state the name(s) accurately? Most of us can improve our listening skills and can avoid an incident such as the following one.

One of the authors recently set out to change the locks on an apartment house since the tenant moved and did not return the keys. The owners lived downstairs and had two outside doors, while the apartment had only one. The locksmith was asked to key the locks so the key for the two doors would also open the apartment and the apartment key would only operate the apartment door. In his haste the locks were reversed, with the apartment key opening all three and the owners' only their two.

The locksmith had to redo the locks on the three apartment doors. This cost him time and materials that effective listening would have saved. The locksmith was already looking at the old key figuring out how to change the pins and reversed the order.

Helping.   Helping people is a satisfying experience for most of us. If our enterprise is helpful, and the costs are competitive, it will be patronized. People appreciate the helpful person who cheerfully recommends a competitor when he or she doesn't have what we want. The extra touch is appreciated, as demonstrated by Renita, Figure 25–2.

Figure 25–2   Polishing a customer's tack earns Renita high marks. (Courtesy Lester Bean)

Performing.   People usually continue to patronize a business that performs well. It is important to live up to our commitments—to deliver when promised and to honor all agreements.

Phil was approached by a large soybean farmer for a bid on his crop of soybeans. Phil obtained details as to the number of acres and the yield. He then contracted the beans to his terminal and quoted a price for the estimated yield. In the process of delivery, the price at the terminal dropped 7 cents per bushel and the farmer delivered 1,250 bushels more than the estimate. Phil paid the bid price for all the beans rather than cut the price on the overage. Doing this aided Phil's professional image. Business cards also help that image, Figure 25-3.

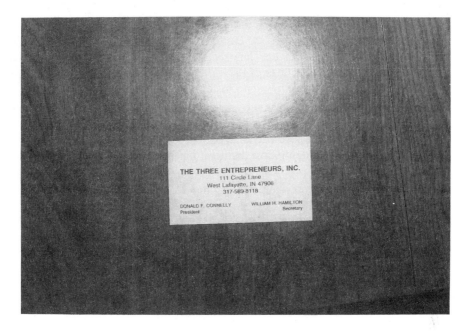

Figure 25-3   Business cards create a professional image.

Serving.   All businesses owe service to the community both to maintain and to improve its quality. Businesspeople who contribute to community development are bettering their home and workplace. Prosperous businesses build traffic for other businesses in the area. The Chamber of Commerce and other business and civic organization members promote community and self-interest through their program participation.

SUMMARY

The skills we have been discussing all involve treating others as you yourself want to be treated.

SELF-EVALUATION

Human Relations Quiz

1. What do we mean by human relations?
2. List the authors' key words in human relations.
3. How do business policies affect relationships with customers?

STUDENT ACTIVITIES

1. Role-play a salesperson who has an unhappy customer on hand. Work with the rest of the class. Enact situations where the customer 1) is insulting; 2) is asking questions; and 3) is very polite. First, have a clerk react as the situation develops. Then have the clerk accept the negative feelings of the customer and show a sincere desire to help. Let the other class members draw conclusions.

2. Role-play the clerk when the customer wants a product not carried by the store. First, have the clerk say, "No, we don't carry that kind of merchandise." Second, have the clerk say, "No, we don't have that on hand but we will be glad to order it." Third, have the clerk say, "We don't have the item, but I believe XXXX has it. Why don't you try them?"

3. Suggest ways business people can serve to improve their communities.

CHAPTER 26

What Should I Consider When Hiring Others?

OBJECTIVE

To identify the major concerns in hiring employees.

COMPETENCIES TO BE DEVELOPED

After completing this chapter you will be able to:
1. List the major concerns in hiring new employees.
2. Identify sources of applicants.
3. Develop a simple job application blank.
4. Identify six items that increase employee job satisfaction.

TERMS TO KNOW

Belligerently	Clients
Compensation	Disgruntled
Financial capability	Individual Retirement Accounts (IRAs)
Job description	Leave of absence
Personnel review	Rank
Resume	Seniority

INTRODUCTION

If you had an unlimited supply of mind, muscle and money, you could handle all of your business transactions exactly as you wished. But your supply of these resources is limited. Right now Renita can handle all the muscle jobs, Figure 26-1. In order to fully satisfy all the mind and muscle needs of her business in the future she, like you, will need additional people. This is the time to hire employees.

Figure 26–1 Renita exercises a horse. (Courtesy Lester Bean)

To a customer, the employee that is dealt with represents the business. Your employees will often be the direct line between your business and your customers. A friendly, courteous, and helpful employee appeals to a customer. Nothing turns away a customer faster than a surly or uncooperative employee.

CASE STUDIES

The Case of the Key Person

Don Connelly, an insurance salesman, was given an opportunity to purchase an existing insurance sales business. This meant that he could buy the building that housed the business, and have the opportunity to serve an established clientele in the agricultural community. The situation offered many advantages for Don, but he understood that in order to maintain these *clients* as his own, he would need to satisfy their insurance needs at least as well as their previous agent had.

Don would need to spend much of each day on the road, selling insurance and adjusting claims. He needed to hire a person to handle the office responsibilities—the paperwork, the filing and the correspondence. More important, this person would often serve as the communication link between Don and his clients. Don knew that this person would be a key to his future success.

Rather than advertising the position, Don spoke with several people in the community about the job. He told them that he was looking for someone with an agricultural background who would understand the needs of his clients, who were mostly farmers.

Maralyn Turner's name was mentioned as someone who might be perfect for the job. Maralyn's resume reflected previous job experience in banking prior to the birth of a child. Her experience included proof operator, bookkeeper, and teller. In other jobs, she acted as an assistant in the county treasurer's office and as a volunteer association secretary-treasurer.

Don contacted Maralyn. Not only had she grown up on a farm, but Maralyn's husband farmed and raised livestock. She was intelligent and personable. She was adept at the clerical demands of the job. Most important, she was sensitive to the needs of the farming community. This, Don felt, was more important than specific training and knowledge in the insurance industry. He hired Maralyn.

She remained with the company for nearly 10 years until the business was sold. Don's instincts about the right person proved to be true. Maralyn was a quick learner and picked up insurance terminology rapidly. After three years in the agency, she attended insurance school and passed the state insurance exam. She then became a greater asset to the business since she could also write insurance for the agency. The office ran smoothly under her guidance. Don's clients appreciated her sensitivity to the particularly seasonal, and sometimes stressful, aspects of farming.

A Case of Miscommunication

Laura Winters had taken a college class called Business Writing, which included an extensive study of *resume* writing. Later, realizing that not everyone has the skills or the training to create a well-written and convincing resume, Laura went into business as a consultant for job-seekers who were willing to pay for a quality resume.

Laura found that she was supplying a service that was in demand in her city. As satisfied clients spoke highly of Laura's contribution to their successful job searches, requests for her service began to outweigh her time. Rather than turn away clients, Laura decided to hire a person to handle some of the tasks that were consuming much of her energy.

After interviewing several applicants, Laura hired Martha Reeves. Martha's duties included taking rough copies of resumes to a local printer who printed the final copies according to Laura's specifications.

One of the reasons for Laura's success was that she always delivered a quality product promptly, often by the next workday. She did this with the cooperation of the printer, who frequently worked a few extra minutes or gave Laura's work extra priority over that of less regular customers. Laura was able to promise her clients timely delivery.

Shortly after Martha began dealing directly with the printer, however, Laura's orders began taking longer to process. Sometimes Martha returned from the printer's empty-handed, saying that he was running behind and wouldn't have them until the next morning. The delays often meant that Laura had to call clients and explain that a piece of work would not be ready when she had promised.

An occasional incident could have been overlooked, but the delays began to happen frequently and Laura decided to speak with the printer. Perhaps she would have to take her business elsewhere.

To her surprise, Laura discovered that the problem was not with the printer but with Martha's attitude. The printer had been giving Laura's work special attention because he had appreciated her friendly and competent business manner. He was honest enough to admit to Laura that Martha rushed into the shop *belligerently,* demanding prompt attention. Laura's orders gradually lost their priority as the printer became weary of Martha's attitude. Laura apologized for her employee's brusque and demanding manner and determined to discuss the issue with Martha.

What the Case Studies Tell Us

- In the first case, Don recognized the importance of finding the right person for his office. To him, personality and background outweighed specific knowledge about insurance since these skills can be acquired. He attributed a great deal of his success in the business to Maralyn's sensitivity to the clients' needs.

- Don asked for referrals from business associates rather than advertising the job he needed to fill. He was able to find the employee he wanted through a reliable recommendation and avoided the time-consuming process of sorting through piles of resumes and interviewing several applicants.

- In the second case, Laura had adequately described Martha's duties. She had, however, failed to communicate the importance of proper attitude. Martha had been pleasant during her employment interview, and Laura had mistakenly assumed that Martha would treat other business associates with the same respect. Communication is the only way to ensure that your employees thoroughly understand what you consider to be important aspects of their jobs.

- Martha represented Laura's business to the printer. He was no longer dealing directly with Laura, and was not inclined to offer special attention to work brought in by Martha. Your employees represent your business to your associates. It is not safe to assume that they will automatically adopt your business manner. If an employee's attitude will directly or indirectly affect your business, then be sure to make that a part of their training.

STEPS IN THE HIRING PROCESS

Preliminary Steps

Finding and hiring qualified employees is not as simple as it seems. You need to do some homework before you begin your first interview.

What exactly are you looking for? What are your needs? Do you want an employee who will perform a specific duty or several duties? A source of employee dissatisfaction is having to perform duties that were not specified in the original agreement.

Not only must you know what you expect of an employee, but you need to know what you, as an employer, have to offer. This might be a good time to review Chapters 20 and 23, which discuss wages, promotions, vacations, time off, grievances, fringe benefits, and retirement policies. These may or may not be negotiable items, but prospective employees will need to know what you will offer them in exchange for their services.

The best way to clarify your policies for yourself and your employees is to have them in writing. You need to develop clear and concise policies regarding each of the following aspects of employment:

Hours. How many hours a week will the employee work? Will he or she be expected to work overtime during peak seasons? If so, how will they be compensated for that? With extra pay? With extra time off when business demands are slack?

Compensation. Will employees receive a wage or a salary? Will you offer commissions? An annual bonus? Incentives? How and when will the payroll be disbursed?

Fringe Benefits. Will employees receive discounts on merchandise? Will your company supply health and/or life insurance benefits? Have you defined a pension plan?

Vacations. How will your employees earn vacation time? Can you be flexible? This may mean doing without a valuable employee during a crucial business peak. Will you expect your employees to take vacations only during the slow season? Will you give vacation pay? Additional time off without pay?

Time Off. What is your policy regarding time off for holidays, emergencies, or personal needs? Will each employee have an allotment of days for unexpected absences? How will you respond if an employee requests a three-month *leave of absence* during a busy time? How will you handle maternity leave for expectant mothers? Expectant fathers?

Training. Will you train new employees, or will that responsibility fall upon supervisors within your management structure? Will you offer a raise at the end of the training period? How will you evaluate the effectiveness of your training procedures?

Retirement. What type of retirement benefits have you planned for your employees? Will you offer an investment plan for those who wish to participate using a portion of their earnings? If so, who will administer this plan?

Grievances. How will you handle conflicts among your employees? Will they be able to discuss unsatisfactory situations with you?

Promotion. Can your employees expect periodic promotion? Will you offer merit increases in wages and benefits, or will all employees receive consistent and periodic raises across the board? Will employees be rewarded with bonuses as the company grows?

Personnel Review. Will you establish a standard procedure for periodic review of employee performance? Will employees be able to discuss their satisfaction or dissatisfaction at this time?

Termination. How will you deal with an employee who consistently gives less than satisfactory work? Will you lay off workers if business is slow? Will employees with *seniority* have priority over newer employees? Will you offer severance pay? What conditions will warrant the discharge of an employee?

Focusing on Your Needs

Your purpose in hiring an employee is to match the skills of an applicant to the needs of the position to be filled. In order to do this, you should first know exactly what you are looking for in a new employee. You need to analyze the job. Once you have the *job description* firmly in mind, make a list of all the skills that will be needed to fulfill the position satisfactorily. You might want to prioritize skills from those that are absolutely necessary to those that would be beneficial but not crucial. As your list grows, the chances that you will find a perfect applicant will decrease. You, therefore, will have to make a judgment about which applicant most closely matches the demands of the job.

Searching for Applicants

Only when you know what type of employee you are looking for do you want to begin the recruiting process. There are many sources for job applicants.

State Employment Offices.   Each state has an employment bureau which keeps files of currently unemployed job applicants. The employment bureau will furnish names of applicants whose skills most closely match the needs of an employer.

Private Employment Agencies.   These agencies, too, will help match their clients' employment skills with an employer's job description. There is a fee for this service, however. It is usually a percentage of the employee's wage or salary over a period of a month or two to a year. This fee may be paid by either the employee or the employer.

Newspaper Advertisements.   "Help Wanted" advertisements in newspapers usually generate a large response. Some employers who do not wish to screen a large number of applicants avoid this method.

Community Schools.   Some employers post job listings with high school guidance counselors, especially for part-time jobs that can be handled by students, or for jobs that do not require advanced education. A good example is babysitting opportunities, Figure 26–2. Nearby technical schools or colleges are also a good source for job applicants.

Figure 26–2   Hiring a babysitter allows the mother to have time for other activities. (Courtesy Susan M. Peters)

Referrals. Some employers contact friends, business associates, or fellow members of the Chamber of Commerce for leads on potential job applicants. A referral offers the assurance that someone they trust can speak for the capabilities of an applicant.

Developing Application Forms

Chapter 23, "How Can I Establish a Sound Set of Business Procedures?", contains a simplified application form (see page 247). You can modify that according to the needs of the job you want to fill. You will want the application detailed enough to answer your questions about the applicant's skills and experience, but simple enough to scan easily. Remember that you, as an employer, are required to stay within some legal boundaries regarding the questions that you are allowed to ask an applicant. (These regulations apply to verbal as well as written questions.)

Civil Rights Act of 1964. This law prohibits employment discrimination because of race, religion, sex, or national origin.

Public Law 90-202. This regulation prohibits discrimination on the basis of age with respect to individuals who are at least 40 but less than 70.

Protection of Physically Handicapped Persons. Federal law prohibits discrimination against persons who are physically handicapped.

An employer should have all applicants fill out a job application as a first step. This allows the employer to screen applications to find the best qualified applicant and can save countless hours spent in interviewing.

Conducting the Interviews

Interviews are time consuming, and job interviewers can feel as nervous about the process as the applicants that are being interviewed.

The purpose of the interview is to allow the employer the opportunity to select from among the most qualified applicants (determined from the applications they have completed). The interviews will flow more easily if you are prepared and have specific questions. "What are your job skills?" "What did you do in your last job?" "Why would you like to work for this company?" "What do you feel you can offer us?"

Keep in mind that you, too, are being evaluated. Participation in a job interview does not mean that the applicant, if offered the job, will decide to accept it. Offer the applicant an opportunity to ask you questions about the position, the benefits that you can offer, and the business itself.

Verifying Information

You have obtained information concerning applicants' previous employment. Double check that information with the employer. How did this person perform on the job? Is their list of duties accurate? Why did the employee leave his or her previous employment? Given the opportunity, would the employer rehire the applicant?

Most job applications request references. Contact those references to inquire about a person's qualifications as an employee, a student, or a volunteer. Keep in mind that personal references are listed because the applicant knows that the referral will be positive.

Selecting an Applicant

When you have devoted many hours to screening applicants, you will narrow the list to the two or three best qualified. You may decide to ask these applicants to return for a second, or even third, interview. Eventually, you will decide on the applicant you want for the job. To elicit the quickest response use the telephone to extend the job offer. In the event that that person has, in the meantime, accepted another job or decided against accepting the job, you can offer the position to one of the other well-qualified applicants. Once you have hired someone, you should immediately notify all other applicants, in writing, that the position has been filled. You may decide to keep the applications that you received for future reference.

Paying Your Staff

In Chapter 20, we discussed the difference between a salary and a wage. It is time to develop a formal plan for paying your employees. In effect, you are rewarding your employees with money and benefits (which equate to money) for services that they render for your business, and you must know exactly what rewards you can offer. The following steps will help you determine just how to develop and begin a pay plan for your staff.

Define the Jobs. It is possible that you will be hiring more than one employee as you begin business. They may or may not be hired to do similar jobs. If you were a building contractor, perhaps you would hire a crew of six people who would work along with you every day, framing a house one week, wiring for electricity the next, and finishing dry wall the third. In this type of business, the job description would change from day to day. Imagine that you were to open and manage a new store. Along with sales staff, you might hire a buyer and an accountant. These jobs require more training and, understandably, command higher pay scales. Whatever your business, you need to define the jobs that are a part of it.

One way to do this is to create a written job description for each position. This could include information such as:

Job title	Experience required
Level of supervision	Beginning pay level
Main duties	Advanced pay levels
Other duties	Maximum pay level
Training required	

Evaluate the Jobs. Job descriptions for each job within your company will help you determine and *rank* the complexity and responsibility that each job entails. This can be an invaluable step in helping you determine the worth of each job to the business, and, therefore, what pay levels can be offered. You can evaluate the jobs according to the contribution each makes to the success of the business. You are actually assigning a value to each position. It is not unusual for a business with 100 jobs to have 10 or 12 different pay levels.

Price the Jobs. Once you have placed a relative value on each job within your business, you need to decide what each job will pay. Pay levels vary greatly across our country, and are established by several factors, from demand to cost of living. The best way to determine the pay scale you will offer is by doing some research into pay levels in the community where you establish your business. There are several sources available to you, including the local Chamber of Commerce or trade associations. The U.S. Bureau of Labor Statistics can provide generalized information about pay levels for jobs in different areas, and your State Employment Agency will have information about job pay in your area. You may even be able to discuss job pay levels with the business manager of a company similar to yours. By consulting several sources, you can get a good idea of the average pay offered for different jobs in your area.

Next you need to determine a pay range for each job. Many employers establish a minimum and a maximum pay level for each job. You may wish to make this flexible in order to handle special situations, such as an employee who comes in with greater than average experience, or the recruit who quickly demonstrates surprising efficiency.

Install the Plan.   Once you have established a pay range for each job, you need to decide how and when you will reward your employees by increasing their pay. It is crucial that you develop a *personnel review* system that is simple and fair. An employee who feels that he or she is unfairly treated will become *disgruntled* and unhappy. Job performance will quickly reflect this dissatisfaction. On the other hand, an employee who feels fairly compensated develops feelings of loyalty and appreciation for the company, and job performance will reflect this.

There are many approaches to administering individual pay increases. Some employers offer automatic raises with time of service. Some offer merit increases, which recognize performance and contribution. Some promote employees to the next higher pay range. Some base wage increases on the inflation index. Your goal is to come up with a combination of these methods that will attract new employees and keep current employees satisfied. Your plan will need to be adjusted according to the *financial capabilities* of your company as well as to the pay levels offered by other employers in your area.

Managing Employee Benefits.   We have discussed direct pay for employees. Benefits for employees have come to represent a substantial cost to the employer, one that can neither legally nor ethically be ignored. According to a 1978 U.S. Chamber of Commerce survey, costs of benefits comprised 31.7% of payroll expenses during that year. Many job applicants base their selection of a job on the benefits offered.

Some benefits, such as social security, workers' compensation, and unemployment insurance are federally mandated. Also, states sometimes require employers to provide temporary disability insurance. Your legal consultant can explain the regulations that will directly apply to your business.

Benefits are generally classified as general or flexible employee benefits.

General Employee Benefits.   General employee benefits are given to all employees. These usually include insurance benefits, such as term life, hospital, major medical, long-term disability, travel accident, dental, and maternity. The employer usually pays the entire premium for the employee, and offers family coverage with the lower group rate at the employee's expense.

Flexible Employee Benefits.   Flexible employee benefits are offered by the employer, often at discount or group rates, but are paid for by the employees. Employees are not obligated to participate in these benefit plans, which may include supplemental life insurance for family members, *individual retirement accounts (IRAs),* and personal accident insurance. Such insurance plans are attractive to employees because, as part of a group, they can purchase such coverage much less expensively than they could individually. When an employee elects to participate in an optional plan, it is usually paid for through payroll check deductions.

Using Temporary Help Services

Imagine a situation in which your business needs to increase its production temporarily. Perhaps you need to make a seasonal adjustment, or a new order taxes the capabilities of your staff. They could handle the extra demands by putting in hours of overtime, but that becomes too expensive for you. You might want to consider hiring temporary help.

Temporary help services exist to supply businesses with skilled workers on a short-term basis. This has several advantages for the employer who needs extra help for a short time but cannot justify hiring another permanent employee. You are actually hiring the services of the firm, rather than the employee. Therefore, you pay the firm (the temporary help service) which in turn pays the employee. The temporary help service, then, is responsible for the employee's payroll, bookkeeping, tax deductions, and benefits. You avoid having to find and train a new employee.

The direct cost, or wage, that you will pay for the use of temporary help is often higher than the going rate for similar jobs. This is because the administration costs and a profit for the service are built into the charge. You must weigh the advantages against the disadvantages, and the cost against the savings, to determine whether the use of temporary help is the best solution in your situation.

If you decide to use temporary help, be prepared so that you can make the best use of the time. Be sure to give the service all the information possible about the job you need filled. Only with adequate information can the service provide the best qualified person for your job. You must also arrange for supervision of the help so that the job will be clearly explained and the work will be closely observed. Inform your employees about the situation and elicit their cooperation.

Evaluate the performance of the temporary helper. If you are not satisfied with the results, you should notify the service firm immediately. Most services will replace the person as quickly as possible, and many do not charge for the unsatisfactory workers if notified within a few hours. Use this experience to help evaluate future situations, in which you may again need to decide whether to hire temporary help.

Employee Satisfaction

We have talked about compensating employees by both pay and benefits. There are many other ways to increase your employees' satisfaction with their jobs and with your company.

An extensive amount of research has gone into the effects of employee satisfaction on business productivity. It almost invariably demonstrates that increased employee motivation and job satisfaction directly result in increased output. The following are a few time-tested management policies that, in addition to adequate monetary *compensation,* can increase employee satisfaction:

- Honest and consistent employee evaluations
- Fair and objective job performance standards
- Encouragement of individual development through training and educational programs
- Open communication with and among employees and receptiveness to positive criticism
- Flexible and diverse work assignments which allow variety and challenge

- Encouragement for employees to become involved in decisions affecting their jobs
- Clean, safe, and healthy working conditions
- Encouragement for employees to tailor their jobs to their skills
- Demonstration of recognition and appreciation for outstanding work

In addition, some companies have found the following incentives to be a great motivation for their employees:

savings programs	parking privileges
recreational programs	legal assistance
discounts	extra vacation
scholarships	child care
loans	job titles
profit sharing	trade association memberships
company car	flexible hours
personal expense account	team assignments

The satisfaction of your employees can generate both savings and profits for your business. Satisfied employees tend to remain with the same company, thereby helping you to avoid costly turnovers which involve time and training. Employees who are happy with their jobs do better work, which means more quality output for your company. Reports from your employees can augment your firm's reputation, which in turn can attract new customers. As you can see, your efforts to increase the satisfaction of your employees are good for all concerned, as you work toward the success of your business.

SELF-EVALUATION

Hiring Others Quiz

True-False: On a separate sheet, write T for true statements and F for false statements.
1. Any employee can create the customer's image of the business.
2. Advertising a position is seldom the best way to find an excellent employee.
3. A valuable employee is one who has a desire to improve.
4. Teaching employees about their job duties should include public relations training.
5. Only employees in the front office need public relations training.
6. Employers sometimes need feedback about their employees.
7. Every employee wants to be the best.
8. Every employee should try to be the best.
9. With the amount of unemployment that exists, the number of potential employees is ample.
10. A good way to determine what you want in an employee is to write a job description that explains the responsibilities.
11. In a small business such things as fringe benefits and vacations are not as important to hiring as in a larger business.

12. Time off is easier to arrange when the business has few employees.

13. The way a small business arranges its pay scale is independent of other businesses in the area.

14. It is impossible for a small business to consider such items as retirement benefits.

15. Personnel review means evaluating people and making employees accountable for their actions.

16. Personnel review means to check the employees' performance for a period of time.

17. Sources of prospective employees used vary greatly in the results produced.

18. Referrals can be an excellent source for potential new employees.

19. Putting the interviewee (person who is applying for the job) at ease is important to allow their real personality to come through.

20. References listed on a job application should be checked.

21. A person who was fired from the last job may become a valuable employee to your business depending on the circumstances.

22. Employees should be fairly compensated for their contribution to the success of your business.

23. With several levels of jobs in a business, pricing them fairly is a difficult task.

24. Temporary help is not as committed as regular employees.

25. Employee satisfaction is of value to any business.

26. Allowing employee input on things that affect their jobs is asking for trouble.

27. Recognizing superior performance builds employee morale.

28. Job incentives pay dividends to the business.

29. Evaluating the performance of temporary helpers is a waste of time.

30. It is always appropriate to treat an employee as you would want to be treated if your roles were reversed.

STUDENT ACTIVITIES

1. With your classmate role-play the employment interview for a common local job.

2. Gather the job applications completed for an earlier chapter and evaluate them for a given job.

CHAPTER 27

How Do I Manage Business Risk?

OBJECTIVE

To identify the risks that occur in business and to recognize risk management strategy.

COMPETENCIES TO BE DEVELOPED

After completing this chapter you will be able to:

1. Define the following key words and concepts:

Insurance	Occurrence
Speculative risk	Coinsurance
Pure risk	Replacement cost
Loss frequency	Actual cash value
Loss severity	

2. Identify four alternatives for managing business risk.

3. Recognize profit and creativity as motivations for starting a business venture.

4. Identify six reasons for business failure due to speculative risk.

5. Explain the economic incentives that are factors motivating and influencing human behavior in risk taking.

6. Explain the economic implications of consumer demands in a modern society.

7. Plan business insurance needs.

TERMS TO KNOW

Accident	Agent
All-risk insurance	Appraiser
Arbitration	Arson
Basic coverage	Benefit package
Capital	Cash settlement

Coinsurance	Commission
Deductible	Economic risk
Economic risk	Equity
Insurable	Loss frequency
Loss severity	Negligence
Premium	Replacement
Risk avoidance	Risk management
Risk reduction	Risk retention

INTRODUCTION

An entrepreneur is a person who organizes, operates, and assumes the risk for business ventures. Risk management is a factor that every successful business enterprise has carefully included in its planning. In this chapter, we will discuss the larger risks involved in setting up a new business, the process of weighing possible risks against potential benefits, and some of the causes of business failure. This chapter includes a self-quiz that will help you evaluate your risk rating. We will talk about some of the methods of transferring risk through different types of insurance. We will look at some of the different reasons that businesspeople decide to take risks in the following case studies.

CASE STUDIES

The Case of the Dead End Job

As a high school student, Bob earned all of his pocket money by mowing lawns for his neighbors. He liked working outdoors, and enjoyed the satisfaction he felt as he looked over a freshly mowed, well-groomed lawn. The father of Bob's high school friend, Jim, owned a lawn care service and offered both boys a job when they graduated. At first, it seemed like a golden opportunity to Bob, and he proved himself to be a good employee. Many of the customers of Edward's Lawn Care complimented Bob's work. He had also made a couple of suggestions for improving the service that had increased the company's profits. But he was becoming dissatisfied with his job. Because Jim's father owned the business, Jim was given more and more responsibilities, while Bob remained the muscle of the enterprise. In addition, Bob had talked with some of the customers about their need for a landscaping service. Bob saw a definite market for landscaping consultation, and suggested adding that service to the business. Neither Jim nor his father, Mr. Edwards, wanted to make any major changes in the business.

The biggest cause of Bob's frustration was his increasing awareness that his position with the company wasn't going to change. He had worked for Edward's Lawn Care for almost eight years. It was likely that Jim would own the business some day, while Bob would remain the trustworthy employee. The $400 he made every week was ample for now, but one day he would need enough to support a family and perhaps buy a house.

Bob began to consider an alternative. A local nursery offered a limited landscaping service, but Bob knew from his contact with customers that he could design landscapes at a more reasonable cost. He had a reputation in the community as a hard worker and the idea of creating landscape designs really appealed to him. Frustration sometimes leads to entrepreneurship, Figure 27-1.

Figure 27-1 Bob's interest in landscaping triggered his entrepreneurship. (Courtesy Donald F. Connelly)

Bob realized that his $2,000 savings wouldn't begin to cover the cost of starting his own business. He would need to hire someone to be his muscle, while he concentrated on selling his ideas to customers and designing lawn plans. He would need to invest in a used pickup to haul shrubbery and sod. He would also need insurance—both workers' compensation and liability.

Bob talked with his Aunt Mary who was excited about his ideas. She agreed to invest $8,000 in his venture. With that, Bob could borrow the additional *capital* he needed from the local bank.

Bob carefully weighed the possibilities. If his business failed, not only would he lose his and Aunt Mary's savings, he would have a large debt to repay, and he had no guarantee that he could find another job that he liked. But he felt sure that there was a need for a landscaping design company in his community, and he had confidence in his ability to provide that service.

Bob decided that his potential success with a new business outweighed the risks he would be taking. He turned in his notice to Edward's Lawn Care and headed for the bank.

Let's look at the changes in the three entrepreneurial resources that Bob will make. Figure 27-2 illustrates these changes. Bob will now use his mind skills in his own business instead of for others. He will also use his muscle for his own enterprise. His efforts will cause money (and profit) to flow into his business instead of someone else's business.

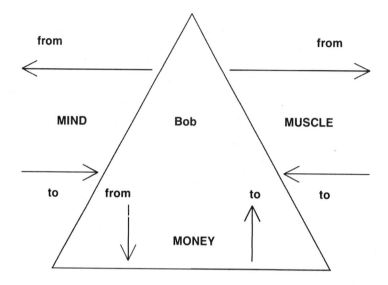

Figure 27–2 Bob's changes in resources.

The Case of the Undesirable Transfer

In Chapter 2 you met Ted and Mary Ruth. You will recall that he had worked successfully for a number of years with an extermination company. His record earned him a promotion that would have moved him to another state and taken him away from the customer contact he preferred.

Ted wasn't given the opportunity to refuse the transfer. If he wanted to continue working for this company, he would have to move. It was Mary Ruth who suggested that Ted set up his own small extermination company. He knew the business well from his experience with the large company. Although they would have to live on a tight budget for a year or two, Mary Ruth's salary as a registered nurse would pay the bills. He could set up an office in his basement, hire a secretary and an employee to work in the field, and continue to devote his time to direct sales. He would have to invest some of their savings to get the business started, and he would have to buy insurance coverage to protect that investment. Ted and Mary Ruth decided that the risks involved in starting a new enterprise were worth taking in order for Ted to continue working at the kind of job that he really liked.

Ted, too, will be affected by the three entrepreneurial resources as illustrated by Figure 27–3.

What the Cases Tell Us

Setting up a new enterprise is a risky business. An entrepreneur puts money and time on the line. Bob's and Ted's experiences raise some questions for us to consider.

- What were the motivating forces behind their decisions?
- What are the risks they take as they begin their business ventures?
- How do they plan to manage those risks?

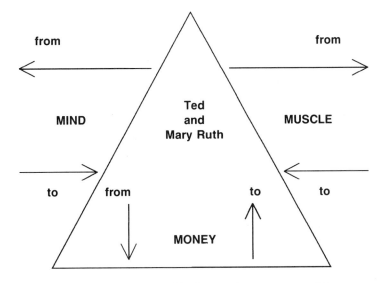

**Figure 27–3 Ted and Mary Ruth's changes in entrepreneurial resources.
(Courtesy The Three Entrepreneurs)**

The case studies illustrate some very important points about decision making and the assumption of risk.

- Bob had two good reasons for his decision. He wanted the opportunity for greater profit in the future, and he wanted to use his creativity to design landscapes.
- In the first case, Bob was no longer content being the muscle of the business. He wanted to become the mind, and in doing so, would invest his money in order to generate larger profits for himself.
- Ted didn't want to relocate, and he wanted to continue working directly with his customers. He preferred small business operations to large.
- In Ted's case, he enjoyed being the muscle of the company. He realized that he would have to invest more of his own mind and money to remain working in the same sales capacity.
- Both Bob and Ted look forward to the personal challenges their decisions present.
- Both Bob and Ted are risking their income, their capital investment, and their job security.
- Both Bob and Ted are assuming responsibility for their new businesses and any liability that might occur.
- Both Bob and Ted realize that they needed to protect themselves and their new businesses by purchasing insurance.

SELF-QUIZ—WHAT IS MY RISK RATING?

The following quiz is designed to give you some idea about your ability to manage risk. It can help you evaluate those areas of *risk management* in which you are skillful. More important, it can help you discover those areas in which you should concentrate to develop the skills you'll need to make wise risk management decisions.

1. You have founded your company on a shoestring. You purchase a 22-year-old jeep for $150 to provide transportation to and from your office in the back of a warehouse. You've bought the required amount of workers' compensation insurance to cover your two employees, but you've reached the limits of your budget. You must choose between a newer model sewing machine which makes a fancy stitch that you think your customers will like, or liability insurance for the old jeep. You decide to: A) Insure the old jeep. B) Buy the better sewing machine.

2. You are scheduling your flight for a cross-country trip. You can either buy a $230 ticket for a major airline's scheduled flight, or you can buy a $145 ticket with an economy airline, which may mean waiting for available seating on a less comfortable flight. You: A) Take the cheaper flight. B) Opt for the scheduled flight.

3. Your own business is doing well, and you have some surplus capital. Your accountant advises that you buy Certificates of Deposit at a 9% interest rate. You suspect that a new advertising campaign would help to expand your business. You: A) Invest in advertising. B) Buy the C.D.s.

MANAGING RISKS

Risk Taking—Success or Failure

In Chapter 1, we discussed briefly the reasons that businesses fail. You need to think about your attitudes toward success, failure, and risk taking before you begin your business enterprise. It's important to become a good judge of risk, but unfortunately, the only way to do so is through experience. Fear of failure may be one of the greatest determents to creative thinking. It is important for an entrepreneur to learn how to improve the chances of success. Sometimes that means dealing with failure along the way. You'll need to put small failures to work for you. By doing that, you can learn what doesn't work, and perhaps find an opportunity for a new approach. Through experience, you can sharpen your judgment of risks. Learn not to make the same mistake a second time. If a promotional campaign is not as successful as you had hoped, try another approach.

We learn early to avoid failure. That works well in the classroom when we're graded on our work, but that sort of caution may not be useful for the entrepreneur who is trying to put a new idea into action. Be prepared to experiment, to try your creative ideas, and to expect some small failures along the way. Consider timing when you want to implement a new idea. Some of the best inventions would have been unsuccessful if introduced at the wrong time.

Every entrepreneur is in a unique situation. If there were a tried and true path to follow for everything, then no one would be an entrepreneur. You learn from the experiences of other entrepreneurs. The following is a list of some of the reasons new ventures fail.

Reasons for Failure

- *Poor market evaluation.* Consider the public when you are deciding whether or not your idea will sell. One of the first economic principles a business manager learns is to judge what the market will bear.

- *Poor planning.* Think ahead. Develop a business plan based on your projected cash flow, and then be prepared to adjust it as the need arises.
- *Underestimating costs.* Talk to others in related fields. Don't hesitate to ask for advice. Make use of the lessons that others have learned as they set up a business. Leave room in your estimates for unforeseen expenses. The only thing you can be sure of is that they will appear.
- *Failure to specify concrete terms.* Your relationship with suppliers is vital. Make specific long-term agreements that will be consistent with your cash flow, and that you can be sure will not change unexpectedly.
- *Lack of support.* A new venture places a great amount of strain not only on the entrepreneur but on family and friends as well. Be sure that those you are counting on for support understand your commitment and are prepared for the pressures that lie ahead.
- *Dishonesty.* It's extremely important to be totally honest with your creditors, your employees, your suppliers, and your customers. Loss of credibility can doom a new enterprise.

Reread the *Wall Street Journal* message, ''Don't Be Afraid to Fail,'' which is reprinted in Chapter 1.

Risk Management

Bob and Ted balanced personal and economic reasons to reach their decisions. There was no way to insure their personal satisfaction, but both knew to transfer some of their *economic risk* by purchasing insurance. Newspapers are full of stories about lawsuits or job-related injuries. Even the best management skills cannot possibly avoid all mishaps, and the possibility exists for that one lawsuit which has the potential to devastate a company. There is protection against that sort of risk. The decisions about how much and what kind of insurance to buy for economic protection constitute what is termed risk management. Figure 27–4 represents risk management in a swine enterprise.

Management Options

By starting a new business enterprise, an entrepreneur creates risk. Excluding the choice of ignoring that risk, there are four different options for risk management.

Risk Avoidance.   To a certain extent, a business can avoid some risks by avoiding activities that are most risky. A roofing company could refuse to shingle steeply pitched roofs on any building more than 30 years old. You can quickly see that the business would be limiting itself, and leaving the door open for the competition. Some *risk avoidance* comes from wise business decisions, but not all risk can be avoided.

Risk Reduction.   Some risks can be reduced. Most companies today take safety precautions, and many offer safety programs for their employees. Most company planners develop alternative strategies for times when projects don't go according to plan or schedule. However, the best *risk reduction* plan cannot avoid every costly catastrophe.

Risk Retention.   All businesses are responsible for at least a part of any risk incurred by the company or its operations. The decision of how much the total *risk retention* should be and how much risk to transfer depends upon, among other things, the available capital that a

**Figure 27–4 Rob practices risk management. He vaccinates the pigs.
(Courtesy Kathy Shanks)**

company has. If emergency funds are unlimited, then the entrepreneur might be willing to assume a greater part of the risk. That, however, leaves less capital for investment and growth, and again limits the potential of the company.

Risk Transfer.   The last option is to transfer the risk, at a known premium, to another agent (such as an insurance company), in order to protect against unknown costs. The cost of transferring the risk is a fixed cost, and one that can be worked into the cash flow of the company. In purchasing insurance, a company is purchasing security against unknown and possibly ruinous expenses.

Insurable Risk

There are two types of risk, speculative risk and pure risk. Speculative risk involves the investment an entrepreneur puts into a new business venture. In the case of business failure, that investment is not *insurable*. Pure risk, on the other hand, is insurable. The process of good risk management involves a series of steps that begins with identifying the risks. Everything, from loss to injury, that could pose a threat to your enterprise should be taken into account. Once the risks have been defined, the manager needs to evaluate each risk according to the probability it will occur and the possible cost to the company. The manager must then decide how much of the risk the company should assume and how much of the risk should be transferred to the insurance company. This can be done by comparing the potential cost of the risk taken and the cost of insuring against that risk to the financial situation of the company.

What is an Insurance Company?

There is a lot of unnecessary mystery surrounding the concept of insurance companies. Insurance companies are in the business for a profit, just as you will be. Insurance companies spend lots of time and money figuring out what risks they can afford to take. If someone insured your building alone against fire, that person would run a great risk of losing money. For a relatively small fee, he or she would be undertaking exactly the same risk that you are transferring away from yourself. However, if an insurance company insures hundreds of clients against fire damage, then the chances that several of those properties will burn is considerably less. Insurance companies have the advantage of dealing in very large numbers. There are formulas to estimate the number of buildings that will burn down in a given time. By issuing enough policies, they can be reasonably sure that the premiums they take in will be greater than the losses they must pay out.

We've discussed insurance as a means of transferring risk. Insurance has other economic effects on our communities.

- It helps make businesses more efficient. If a business is uninsured, it must retain part of its capital to cover possible damages. With that risk out of the way, businesses have more capital to invest in themselves for growth.
- Insurance helps keep the price of products and services low. If it were not for insurance, producers would have to increase the cost of their products to cover possible risks.
- Insurance provides assurance to lenders. No bank will loan you money for a building that might burn down next week, making their money impossible to recover.
- Insurance is a means of savings. Some insurance policies build *equity* for the owner of the policy, and can be cashed in for the value of the equity.

Types of Insurance. The following is a brief discussion of three types of insurance that a solid risk management program should consider. The three categories are liability, life and health, and property insurance.

Liability Insurance. Business liability means legal responsibility. The owner of a business is legally responsible for the company, the buildings, the vehicles, and the products. Any *accident* that occurs resulting from improper management of these assets is called *negligence*. Negligence is often ruled in cases where the customer's or employee's carelessness contributes to the accident. There are several types of liability insurance that the entrepreneur needs to understand. Automobile or product liability insurance may be needed. Legally, if he or she employs workers, the business owner must carry workers' compensation insurance. The owner will also want to insure the business property for liability. Collision insurance provides some of Ronnie's work, Figure 27–5.

Figure 27–5 Ronnie's business includes property damage insurance repairs. (Courtesy Ronnie Orem)

The cost of insurance depends in part on the amount of the *deductible*. As the initial expense that the business is willing to withstand increases, the *premium* decreases. The entrepreneur may decide that the business could afford the first $1,000 in the event of an unexpected accident, but that any cost above that would hurt the company beyond the point of recovery. It is important, then, to insure for any liability expenses over that amount. That insurance policy would cost less than one with a $200 deductible.

In addition to individual types of liability insurance, there are also all-inclusive liability policies that are designed to cover exceptionally large liabilities over and above *basic coverage*. As the needs for protection vary in different types of businesses, so do types of insurance policies. It is important to work closely with a good insurance *agent* who can explain the different types of liability insurance, what they offer, and their costs.

Health and Life Insurance.   We often hear the term "company benefits." Indeed, that is part of the foundation of labor unions. Most companies today offer some type of insurance program for their employees, and, to be competitive, an employer must offer an attractive *benefit package* which includes life and health insurance plans. Group insurance plans for a company are often less expensive than individual plans. Again, there are many types of policies available, and with the help of a knowledgeable insurance agent, a plan can be adapted that will fit the financial situation of the company while still meeting the needs of its employees.

Property Insurance. Property insurance protects against property loss which may occur due to theft, fire, flood, vandalism, or collision. There are individual policies available for different types of needs. There is also a type of insurance called comprehensive property insurance, which covers most types of property loss with specific exclusions. This *all-risk insurance* policy is usually less expensive than several independent policies, and claims are settled more easily when fewer insurance companies are involved. Certain company assets, such as automobiles, jewelry, or machinery, may be specifically excluded from a comprehensive policy. In this case, additional individual coverage will be necessary.

There are many different types of specialty insurance that can be purchased for uncommon needs. Musicians insure their hands, and dancers, their legs. While most businesses wouldn't be interested in these types of policies, they demonstrate the wide variety of coverage available. A business can insure against interruption in production so that it can continue paying wages while its income is temporarily reduced. With credit insurance, the company receives payment for its products or services in the event that a customer cannot complete payment. If an office complex includes large and expensive plate glass windows, a manager may decide to carry insurance against glass breakage. There are insurance policies available, at a price, for nearly any occurrence. The goal of good risk management is to balance the security needs of the company against the cost of that security, and to see that the company is not forced to close because of a financial tragedy.

Selecting an Insurance Company

A person or a business buys insurance for two reasons, protection and peace of mind. There are steps to follow that can help you decide how to buy insurance.

You'll need to decide whether to deal with a direct writer or an independent insurance agent. A direct writer works on a *commission* basis for a particular insurance company. He or she can offer less expensive rates, and will probably have specialized knowledge about the type of insurance offered. An independent agent may represent several different insurance companies. While not specializing in any one kind of insurance, the agent will have a general knowledge of many types, and you may be able to purchase several different policies this way. Your decision will depend on your insurance needs. Perhaps you'll decide to purchase your health insurance from a direct writer, and your liability and property insurance from an independent agent.

Selecting an insurance agent is similar to choosing a family doctor. Remember that the future health of your company may one day rest on decisions that you and your agent make now. You'll want to deal with a person that you can trust. You might want to ask for recommendations from people who manage businesses similar to the one you're proposing. An agent might be willing to give you the names of other local businesses as references. Don't hesitate to contact them. Ask how quickly and satisfactorily claims they have made were settled.

There are other factors you'll want to keep in mind as you select insurance.

Cost.   The cost of insurance coverage may vary widely from company to company. Don't be afraid to shop around. Comparisons may be tricky. While one company may offer lower premiums, a second may pay dividends. Another may appear to be the cheapest, but may have a history of inadequate claim settlements.

Flexibility. If a company doesn't have a policy structured for your needs, is it willing to adjust the coverage? If your business has an unusual need, it might be less expensive to purchase a rider for an existing policy than to buy broad coverage that gives you more protection than you need.

Stability. Don't put your company's protection in the hands of an insurance company that is not financially secure. *Best's Insurance Reports* is a publication that rates the reliability and stability of insurance companies. Poor insurance coverage isn't one of the risks you should take.

Coverage. One of the advantages of working with an independent insurance agent is that you can avoid overlapping coverage. You don't want to pay premiums on two policies that include coverage for the same thing, particularly when most policies state that payment will not be made for claims that are covered by other insurance. You may find yourself in the middle of a debate between two companies, neither wanting to pay your claim.

Remember that an agreement with an insurance company is not a lifelong commitment. Evaluate the service of the insurance company or your agent after a reasonable period of time. Are you satisfied with your relationship and the service you receive? If not, don't hesitate to talk with a different agent.

What is coinsurance? Sometimes an insurance company will require you to insure a certain percentage of your property's value. They do this to protect themselves against expensive losses. In return for your cooperation, they'll offer you a reduced rate on your premium. When you turn in an insurance claim for damage your business suffered, the insurance pays you according to the percentage rate your *coinsurance* clause specifies. If you were required to carry insurance on 80% of the value of your property, and you did so, then your insurance would pay the full amount of your loss, up to the insured amount. If, however, your property had increased in value, and you had not increased your insurance accordingly, then the insurance company would reimburse you for only the percentage of your coinsurance coverage.

How much deductible? The deductible amount of an insurance policy is that amount that you will pay before the insurance coverage begins. As the deductible increases, the cost of the insurance decreases. For instance, if you want to insure your buildings for $80,000, a company may offer three alternatives. If they are to pay all, or a certain percent, of losses over $500, you annual premium may be $1,500. If you are willing to withstand the first $1,000 of a loss, your premium may be reduced to $1,200. However, if your company is in a position to cover the first $2,000 of a loss, the premium may be as low as $800. That's an annual savings of $700 over the cost of the $500 deductible policy. However, if you suspect that your new business could not afford to pay the first $2,000 in the event of a loss, you will feel safer with the lower deductible. Again, the decision depends on the financial situation of your company and the amount of risk you want to assume.

Setting Losses

Your insurance responsibilities do not end with the purchase of a policy. In the event of a loss, there are certain steps that you have to take to make a claim.

Notice of loss. You must inform your insurance agent as soon as possible after a loss. A delay may complicate your ability to collect the insurance settlement. Also, time is a factor when an insurance company needs to investigate the reasons for the claim, as in the case of arson.

Proof of loss.   An insurance company will need proof that a loss has occurred. It's important that you keep accurate records to show that the item in question was a part of your business and has suffered damage. Your records will need to include an inventory of your assets, receipts, and other documentation. In the event of an accident or illness, you'll need medical records.

Selecting your options.   If not spelled out in your policy, an insurance company may give you the choice of a *cash settlement,* repair, or *replacement.*

Arbitration.   In the event of a small claim, your agent may make the settlement directly to you. For larger losses, many companies send a representative to evaluate the damages. If your estimates are not comparable to those of the *appraiser,* you may need to deal with an arbitrator. In *arbitration,* you and the company will agree to present your cases to an independent appraiser, and abide by the decision.

Potential Problems

Like any expense, insurance rates are subject to increases. The most common reasons for this, beyond *economic inflation,* are *loss frequency* and *loss severity.*

Loss frequency.   If an insurance company is insuring a client who makes frequent claims, it will probably decide to increase the client's premium. In extreme cases, an insurance company may even refuse to continue insuring a client. At that point, the client may find it difficult to obtain any insurance except at extremely high rates.

Loss severity.   In the event of a catastrophic loss which proves to be expensive, an insurance company may decide not to risk a subsequent loss with that client without a substantial premium increase. If the insurance company feels that a business may have been partially at fault for a loss, they may refuse further insurance. It is good to keep an open and honest relationship with anyone you do business with, including your insurance agent.

SUMMARY

This chapter was designed to help you think about the risks your company will face and the ways you can manage that risk. An entrepreneur must assume all of the speculative risk involved in the venture, but some of the pure risk can be transferred through insurance.

We hope you understand the different types of insurance available. You'll want to do some research to come up with the best rates and the best companies for your situation. You should have a better understanding of what insurance companies are, why they exist, and what benefits they offer a business community.

An entrepreneur takes risks that others are not willing to take. A successful entrepreneur knows how to manage those risks.

SELF-EVALUATION

Risk Management Quiz

Multiple Choice: On a separate sheet, write the letter representing the best answer for each of the following questions.

1. Risk management refers to the way a business decides to
    a. retain, reduce, or transfer risk.
    b. limit business liability.
    c. eliminate risk.
    d. spread risk.

2. Insurance companies are able to provide risk coverage to individuals and businesses because
    a. risk is spread over all businesses or people.
    b. experience tells how much loss can be expected.
    c. it is impossible to predict losses.
    d. all of the above answers.

3. Many businesses are never started because the person with the idea
    a. lacks motivation.
    b. does not know how to manage a business.
    c. fears a failure.
    d. lacks encouragement.

4. Many businesses have been started when someone with _____ has been stuck in a dead end job.
    a. adequate finances
    b. good credit
    c. an entrepreneurial attitude
    d. a supportive family

5. Another spark that ignites the idea may be
    a. a demanding boss.
    b. a nagging wife.
    c. a desired transfer.
    d. an undesired transfer.

6. All risks are
    a. avoidable.
    b. transferable.
    c. reducible.
    d. predictable.

7. Insurance coverage should be selected on the basis of
    a. cost alone from the agent.
    b. service alone from the agent.
    c. reputation of the agent.
    d. the agent and company reputations.

8. The company with few employees
    a. needs less insurance than larger companies.
    b. needs less coverage than a larger company.
    c. will always pay out less for insurance than a company with more employees.
    d. all of these answers are incorrect.

9. Selecting the deductible in an insurance policy is based on
    a. the balance between cost and protection.
    b. personal preference.
    c. the agent's advice.
    d. the company selected.

10. Coinsurance refers to
    a. a policy with two companies involved.
    b. risk you are required to retain.
    c. a policy written by two agents.
    d. risk you cover by a second policy.

STUDENT ACTIVITIES

1. Research the question, "How many businesses started in 1989, 1990, and so on?" Make a table showing the results.

2. Research the number of businesses that failed in given years.

3. Determine how many of each year's failures over the past five years were in their first, second, third, or fourth year of business when they failed.

4. Discuss your business plan with an insurance agent to determine the best insurance strategy for your business.

CHAPTER 28

![black bar]

Is There A Basis for Profitable Computer Usage In My Enterprise?

OBJECTIVE

To identify the place of computers in the business plan.

COMPETENCIES TO BE DEVELOPED

After completing this chapter you will be able to:

1. Evaluate the strengths of software programs given an adequate catalog description or demonstration.
2. Recommend a hardware system given the tasks desired, catalogs and specifications.
3. Identify situations suitable for computerization given the necessary background information.
4. Use a word-processing package to produce a document pertaining to your business plan.
5. Identify the type of printer used to print a document.
6. Identify local sources of computer hardware, software, and support.

TERMS TO KNOW

Data terminal	Graphics
Hard disk	Hardware
Input	Modem
Network	Operating system
Output	Printer
Software	Support
System	

INTRODUCTION

Deciding when it is practical for your business to enter the computer age is a major business decision. It will change how you do most of your paperwork. There are many advantages to the use of computers in a business operation. They come at a cost—a cost in terms of learning time for employees and customers as well as in capital funds.

The decision to buy a computer will vary with each business. In this chapter we will attempt to point out some of the decision-making factors. Also, we will point out some advantages of computer use and present a guide for choosing the right type of system.

CASE STUDIES

The Case of the Computer Committee

Early in the personal computer mania, Betty's organization decided they should have a computer to keep membership and financial records and to help publish the newsletter as well as print the letters required for the officers of the association. Since there were several hundred members, a computer seemed an excellent way to reduce the workload on the three office workers who provided secretarial service, financial bookkeeping, membership records, and publishing and printing.

A committee was formed to research the purchase. They decided on a system with a large memory and networked *data terminals* in the other offices. Two types of *printers* completed the *hardware* system.

Although the system design was excellent and adequate, several mistakes were made along the way. First, the purchase was made while the computer *operating system* was still under development so there was no track record to study. Second, the office workers were not included in the process. Third, no computer training was provided. A knowledgeable organization member volunteered to train the people whenever he found time. A fourth mistake was made in not selecting the software needed to perform the desired tasks, or purchasing the equipment that would run the software. The computer selected should run the desired software, Figure 28-1.

The Case of the Dust Collector

Jack had operated a small but profitable bakery for a number of years and was comfortable with his local trade and with the two employees who had been with him for most of those years.

The local computer store owner convinced Jack he needed a computer, then sold him a modest personal computer and *software.* Jack tried to learn the operation but soon found it so frustrating that he gave up and stayed with the system he had developed over the years. He had visual control over his inventory, long experience in keeping up with supplies ordered, a knowledge of customers' needs, and did not find the computer worth the time it took to bring it on-line as a marketing tool.

Figure 28-1   Computer hardware should be selected to operate the desired software. (Courtesy Administrative Business Consultants, Inc.)

What the Cases Tell Us

- Computer purchases should start by determining the jobs that are to be done. Select the software first.
- Involvement of the people who will operate the equipment is desirable.
- The capacity of the *system* should reflect realistic objectives for the system's use.
- Unless people are dedicated to learning to use the system, the investment may be money wasted.
- Some employees will leave the job before they will learn the new system.
- Computers will only work as well as the people who operate them; they must have knowledgeable operators.
- Computers have great tolerance for repetitive tasks and, when correctly operated, can do intricate calculations in seconds.
- Computers provide the opportunity to write, edit, and print multiple copies of a document, thus providing correction-free letters and other work.
- Computers with adequate storage capabilities can maintain easily accessible records with up-to-the-minute reports possible.
- Computers are available in sizes ranging from small standalone models to models capable of handling a number of data terminals, each working on different tasks at the same time.
- Learning to operate a computer involves a personal commitment of time and practice. No one can do it for you.

- A computer cannot help your business unless you operate it with appropriate programming or software.

BUSINESS AND COMPUTERS

Some Questions to Answer

Below are some questions to answer as you consider the advisability of purchasing a computer for your business.

1. Does your business require frequent calculations (such as payroll)?
2. Does the business have a large number of items to keep inventory or other records on?
3. Does the business have reports to prepare on a regular basis?
4. Are projections and current status reports frequently needed?
5. Are word-processing abilities useful to your business?
6. Are charts and *graphics* a frequent need?
7. What is your attitude toward computer use?
8. What are your budgetary constraints?
9. How does one make an intelligent computer purchase?
10. Does any nearby school or college offer a business computing course?

Figure 28–2   Mike inserts a floppy disk to enable the computer to perform the desired tasks. (Courtesy Administrative Business Consultants, Inc.)

Computers are Number Crunchers.   Computers have tremendous capacities for handling both simple and complex calculations, doing in seconds what would take us many hours to do. The computer has the ability to repeat the same calculation routines endlessly, where humans would tend to become bored and consequently become error prone. Thus, if repetitious calculations are needed, a computer can save time and boredom. Mike's computer uses accounting software, Figure 28–2.

What Jobs Could a Computer Do for You?   It is important to carefully consider the types of jobs done in your business that a computer could do faster when deciding whether to invest in a computer. As the personal computer age matures, more and more software is available to complete most common business tasks, regardless of the type of equipment purchased. It would be well at this stage to consult with a person who knows both computers and your type of business. A consulting fee would be a wise investment compared to making an unwise purchase.

How Do You Feel About Computers?   Your attitude toward computers will be an important decision-making point. If you feel comfortable with them, you will be more likely to learn to make full use of them. If, on the other hand, you are uncomfortable with computers, you will be less likely to explore their potential capabilities. Even if you are reluctant to learn, you may find a receptive employee who would manage the changeover and teach you those operations you need to know. Bob felt computers could help his business, Figure 28–3.

Computer Costs.   Your budget may dictate how extensive a system you can install. The way to maximize potential will be discussed later under the section on how to make a computer

Figure 28–3   Bob started with a muffin tin for financial management and now has a complete computer system. (Courtesy Robert L. Maudlin)

purchase. Do not let a package price influence you only because it seems a good buy. There are systems in all price ranges but care must be taken to select equipment to do the desired job.

How to Go About a Computer Purchase

Many books have been written on the subject of computer selection and probably one is available in your school library. Your librarian can help you locate articles on the subject if you want to read more than this brief summary.

The most advised first step in computer selection is that of selecting software that is capable of doing the business tasks desired. Few of us have the programming skills or the time to write and debug the programs needed to do the necessary tasks. Business programs are available to do common tasks. Many integrated software packages are available that perform several different tasks in the same package.

After choosing the packages that will do the jobs you want done, choose a computer that will operate this software. The choice of computer hardware is extensive and many computers now have the capability of operating the same software. Many software packages are also written to operate on various computers.

More important than the brand name of the hardware, consider the comfort and convenience of the keyboard, the monitor size, monitor color, columns on the screen, and so on. The printer driver capabilities and graphics capabilities are also important considerations.

Within the computer selected, the size of memory is a cost factor as well as a capability measure. Many software packages require certain internal memory capacities as well as disk drives. In general, the more memory available, the more sophisticated and user-friendly software can be used.

An important consideration in hardware purchase is the expandability of the computer. Many can now be hooked up to huge memory units called *hard disks* to provide the desired capabilities, or, the hard drive can be installed in the computer itself. Hard drives eliminate the necessity of changing disks frequently. Disk drives provide a means of inputting programs and storing desired data. These are available in various sizes and memory capacity.

In purchasing a printer, one must decide whether print quality is the prime consideration or if graphics are desired. A letter quality printer uses typefaces to print quality letters like a typewriter. A dot matrix printer makes its print by using various numbers of dots, while higher quality printers use more pins in their print heads. In another type of printer, ink jets are used to produce the image, which provides the capability of multi-colored printing. The latest printer is the laser printer, which has many different typefaces and sizes that can be changed as easily as changing a module or cartridge.

The printer must be compatible with the computer's printer drive whether it is parallel or serial. Another consideration in selecting the printer is the print buffer. If the printer has a large capacity buffer, the printer stores the print instructions and content in the buffer while printing. This releases the computer to continue other tasks instead of waiting for the completion of the print routine.

In selecting hardware as well as software, one should consider the *support* available to you as a customer of the computer store. A good local business may be worth a few dollars more compared to the mail-order discount operation with limited or no service.

Consider Taking a Computer Course.   Some colleges are now offering courses such as, "Computerizing a Small Business." If a school near you offers such a course, the investment in time and fees may help you to make an intelligent decision. Tell the instructors of this course of your interest. They can tell you if their course fits your need. Taking such a course would enable most of us to devote the practice time needed to accomplish our learning goal. This author purchased an advanced computer but would go back to the known system when he had a deadline to meet. Preparing this text provided the incentive to learn the advanced system.

Information or Data Networks.   In the process of deciding on the suitability of a computer purchase, one should consider the many data bases that can provide up-to-the-minute information. A farmer can join the AgriData Network and obtain current market and weather information, valuable in making management decisions. This is only one of the many data bases available to computer operators. These networks are also able to transmit electronic mail or messages to other network users.

A *modem* allows your computer to communicate with such networks through telephone lines. In theory any other computer with a modem is accessible in the same manner.

Computers Can Monitor Many Operations.   Computers can monitor many operations at a nominal cost. The heat in your school is likely monitored by a computer which turns the system up in time for classes and down when classes are over. They can even keep track of the day of the week. Computers can monitor the level of light in your parking lot. With the aid of a photoelectric cell, they come on at dusk and are turned off at preprogrammed times.

In the final analysis, one must determine if the investment will increase the decision-making information pool, decrease operator time, or increase productivity enough to pay the costs incurred.

SELF-EVALUATION

Computer Usage Quiz

Multiple Choice: On a separate sheet, write the letter that represents the best answer to the question.

1. Computer network services are best as
    a. sources of current information.
    b. storage places for important documents.
    c. a substitute for a hard disk.
    d. a substitute for program disks.

2. A print buffer is a method of
    a. speeding the printer operation.
    b. speeding document preparation.
    c. increasing the computer's memory.
    d. printing graphics and charts.

3. Taking a computer course can be
    a. a good method of selecting software.
    b. a good way to meet other computer users.
    c. a good way to learn about computers.
    d. a chance to use computers.

4. Print quality in a dot matrix printer is a function of the
   a. proper hook-up.
   b. proper voltage.
   c. computer's printer driver.
   d. number of pins in the print head.

5. In selecting computer hardware, the first step is
   a. identifying price.
   b. identifying a local dealer.
   c. identifying the tasks you want it to do.
   d. identifying the software to purchase.

6. The second step in selecting computer hardware is to
   a. identify its expandability.
   b. identify its compatibility.
   c. identify what programs it can operate.
   d. identify the software you want to buy and use.

7. The computer monitor you select should provide
   a. eye comfort and readability.
   b. a large screen size.
   c. graphics display.
   d. graphics print capability.

8. Most authorities suggest that brand name is less important than
   a. the viewing angle.
   b. keyboard comfort and convenience.
   c. the printer interface.
   d. the print buffer.

9. How you feel about computers is
   a. important in brand selection.
   b. important in printer selection.
   c. important in considering a purchase at all.
   d. not important.

10. A telephone modem
    a. makes your computer accessible to any other computer in the nation.
    b. makes it possible for you to access all other computers.
    c. lets two computers with different operating systems share data.
    d. requires a telephone line for use.

STUDENT ACTIVITIES

1. Create a logo and letterhead for your proposed business.
2. Use a spreadsheet to work out your projected cash flow.
3. Select suitable software for given tasks.
4. Visit a small business that uses computers to see how they are used.

CHAPTER 29

How Do I Manage Business Growth?

OBJECTIVE

To identify the concerns of an expanding business.

COMPETENCIES TO BE DEVELOPED

After completing this chapter you will be able to:
1. Identify expansion possibilities for production, service, or sales businesses.
2. Describe how an expansion affects a business plan.

TERMS TO KNOW

Accounts receivable	Balance point
Block man	Inventory
Market research	Outside salesman
Price concession	Seasonal items
Supply company	

INTRODUCTION

Should I Expand the Business? The decision of when and how to expand a business may be a major financial decision with long-term consequences for the profitability of a business enterprise. In this chapter we will try to present some of the concerns to be examined in making such a decision.

CASE STUDIES

The Case of Related Lines

Phil's feed franchise was one of the major corporations in the feed manufacturing business. One of their major strengths was their complete management program with related products. Their line, however, did not include major items such as hog and cattle feeders.

Soon after Phil opened the Exeter Feed and Grain doors, a customer asked for a hog waterer. Phil wisely told him he did not have any in stock but would be willing to order one and allow a 10% discount for the delay. A quick conference with Ted and Howard led to the decision to stock the quality brand in the field. Since no one in Exeter handled the line, they were able to order the items without the necessity of a minimum order.

This made Phil consider if other items should be added to the *inventory*. The answer at that time was easy—their inventory used up the available capital so further expansion was impossible. After a few months and increased sales, Phil began to find out what items he could sell at a profit.

The Case of the Outside Salesman

Phil's *block man* had been pushing the idea of expansion by means of adding an *outside salesman*. This person would spend all week in the country calling on farmers and selling their line of feed and sanitation products. Since most of the company's capital was invested in *accounts receivable* and borrowed money was used to augment cash flow, Phil could not find the available money to hire a second employee. (He had only one at that time.)

Both of his partners had attended the meeting and thought the idea a good one. They came up with an alternative plan for a trial basis. Phil, who had called on farmers in the area for three years as the local vocational agriculture teacher, would spend his weekday mornings doing the outside sales job. Ted and Howard would fill in at the Feed and Grain store.

The trial period failed to provide the extra cash flow needed to carry the basic salary for an added employee. As a result, this method of expansion was dropped at that time.

The Case of Relocation and Expansion

A small hardware store was well liked by the local population as it carried many useful items. However, it was a jumble of overcrowded aisles and wall shelves, and customers requested many items not stocked because of space. There was also difficulty finding parking in the downtown area.

After thoughtful consideration, the owners decided to move to a new site on the city's bypass where they not only built a new building with several times the old floor space, but also expanded the lines of merchandise previously stocked. The move proved to be a good business decision as adequate parking has greatly increased store traffic. Figure 29–1 shows business growth.

What the Cases Tell Us

- There are several different ways to expand a business: increase staff, increase merchandise or production, increase sales, or increase services.

- Increases or expansion create a new *balance point* or potential for profit.
- Customers are a good source of information on profitable expansion—if you know what customers want and provide it, they are more likely to buy and thus provide you with a profit.
- You must decide on quality versus price in expanding to new lines.
- *Price concession* for out-of-stock items is often a worthwhile sacrifice.
- *Market research* on related items should be continuous and up-to-date.
- What the *supply company* wants you to do may or may not always be in your best interests.
- A short-term trial is often insufficient to effectively test an idea. For example, *seasonal items* may only sell at one time of the year.
- Cash flow can present road blocks.
- Sometimes the only avenue for expansion is through a complete change in the business, such as location.

Figure 29–1   The growth of this business is apparent. They are adding on to the building.

WHEN SHOULD YOU EXPAND?

Considerations for Expansion

We want to examine expansion in each type of business described earlier—production, service, or sales of services or products made by others. We will also look at increased sales which make expansion possible.

Increasing production may be illustrated by the following examples. Mrs. Emma Jeeter was known in her neighborhood for the quality of her potato salad. Her friend who ran a deli requested a supply of the salad and offered her a fair competition price. Soon Emma was making salad for a regular group of customers. As the years went by it became obvious to Emma that her other costs were making the business less profitable. In order to survive the competition she would have to expand her volume or fall by the wayside. She chose to expand and hired a sales and promotional employee who was experienced in other lines. She was hoping to rebuild her volume to the desired level of profit.

Another example of expansion by way of increased production would be Kermit's success in pork production. He first expanded to four sows and litters, then rented a vacant farmstead and expanded to 16 sows furrowing twice a year.

Figure 29–2   When Clint started his business, he used this push mower. He now needs the riding model pictured in Chapter 1. (Courtesy Donald F. Connelly)

In each of these cases the law of supply and demand is in effect. If every pig producer expands, Kermit's profitable feeding margin will disappear. In the case of Emma, the drive to increase her share of the market will bring this law into play. She will receive competition on the level of price, convenience, or quality. Price competition may be illustrated by a business that has more modern equipment, allowing lower production costs. This, then, allows the business to compete successfully with lower prices. Emma believed her superior quality would bring the increased sales volume she needed to regain her desired income level. This is an example of competing on quality. Competition on the basis of convenience can be illustrated by the entrepreneur who decides that individual servings of catsup and mustard or half-and-half would sell well based on restaurants' coffee and sandwich trades.

Expansion in service industries also involves the law of supply and demand since it applies to all economic activity. In the case of services, expansion can be due to population growth, to increased share of the market, or to superior service as recognized by users. Clint's service required a larger capacity mower, Figure 29–2. Renita's expansion occurred through carrying out more of the same activity, Figure 29–3.

**Figure 29–3 Renita expanded her business by boarding other people's horses.
(Courtesy Lester Bean)**

In the case of the production of a product, increases are possible through various means—increasing the time devoted to production, speeding up the processes, increasing the number of machines and operators, or increasing the capacity of each machine. A second or third plant, duplicating the existing processes or updating to newer technology, could also be established.

In the case of most service activities, expansion involves more hours or more operators. Specific services, of course, have differing potential for expansion. Sometimes the additional sales volume can be handled by more efficient operations, such as the drive-through windows at fast-food restaurants.

An example of sales expansion is the use of toll-free numbers to promote call-ins for new orders or convenience for people reordering.

Another concern when thinking of expanding a business is, "What will expansion do to me personally?" Will the expansion lead to what I want most? Will life be richer for me? What will the expansion do to those who I care about, my family, and my present employees?

What will expansion do to the business in terms of size or organization? Will new employees fit in? Will I have to share management? Am I willing to give up some of the control I have on the business now?

An expansion can signal a change from a small closely held business to one with many owners such as a corporation. Those concerns are treated in another chapter.

An expansion may make the management task more than one person can handle. In order to split or divide the new larger job, you should look for a person you can work with comfortably and whose skills complement yours. If you take on a person with great ability and no experience, you will need to train that person. That opportunity is both positive and negative—positive in that you can set the style, and negative in that it can be time-consuming when you are overloaded with the details of the expansion. You may want to start training before expansion occurs so you have help in the transition.

When you hire an experienced person, the training period is decreased and you have immediate help. One disadvantage is that the person's management style is already set and may be hard to change.

Expansion may also be the fuel that drives the entrepreneurial enthusiasm which may have gone stale in a smoothly running business. In an earlier chapter, we mentioned the developers who build businesses started by others. Frequently they get their opportunity when the original entrepreneur tires of a business. Sometimes entrepreneurs with good ideas are not the most effective managers.

You Will Need a New Business Plan

Once you have determined the most logical expansion route, you need to project the costs and returns possible with the proposed expansion. Again we return to a business plan reflecting the planned changes. If borrowed capital is needed, this projection will be an important item in making a loan request. Even if no loan is required, the business plan will let you know if the expansion is indeed a good idea.

Developing the Business Plan.   In order to plan effectively for expansion, one needs to rethink the business plan steps made in the beginning of the business. Special note needs to be made of the conditions leading to the expansion move:

- The capabilities of the present site.
- Changes anticipated to improve the operation.
- What facilities will be needed?

- What equipment will be needed?
- What personnel will be needed?
- How the changes will improve operations.
- How sales will be increased or efficiency increased to improve the competitive position of your business.
- Can changes be made without interrupting the business operation?
- Will local or state approval be needed for the changes?

How you answer these crucial questions may determine your profitability, or even your survival as a business.

SELF-EVALUATION

Expansion Quiz

1. Describe how a business plan needs changing when expansion occurs.
2. Describe how expansion varies by production, service, and sales businesses.

STUDENT ACTIVITIES

1. Locate businesses that have recently expanded and study how they managed the expansion.
2. Write a revised business plan to show how your business can be expanded.

CHAPTER 30

Should I Change the Business Structure?

OBJECTIVE

To examine the effects of changing the business structure.

COMPETENCIES TO BE DEVELOPED

After completing this chapter you will be able to:
1. Describe the advantages of a change in the business structure.
2. Identify structural changes that can limit personal liability.

TERMS TO KNOW

Business perpetuation	Business structure
Employee incentives	Estate
Incentives	Limitation of personal liability
Operating loans	Perpetuation
Personal liability	

INTRODUCTION

In one of the earlier chapters we examined the characteristics of various types of business organizations. In this chapter we plan to look at the idea of changing the structure of a business, the reasons for such changes, and some of their possible consequences. The way these changes are accomplished will complete the study.

CASE STUDIES

The Case of a Change in Partners

After the Exeter Feed and Grain business had been operating about two years, Howard came to Phil and talked about the possibility of his leaving the partnership. Howard had just invested in a business his son-in-law was interested in and this made his wife uneasy. Howard had lost all his investment capital once and it had been a hard battle back to a relatively strong financial position. Howard's wife had no intentions of going through this a second time and asked him to limit his liabilities. Since the new business was a corporation, the partnership in Exeter Feed and Grain was the main point of exposure that could be removed by an agreeable sale. Phil understood and agreed to a change in the unlimited partnership structure.

A Case of Divided Assets

The Carelton family had prospered in their farming community and when the oldest son graduated from college with an animal science degree, he was welcomed home to take over the family dairy business. In searching for an equitable way to manage the change, the family lawyer suggested the formation of a corporation.

The family had five children so the business was set up with the parents owning 60 percent of the stock and the children 8 percent each. Wages were paid and stock purchase means established to permit growth in ownership percentage and in business net worth.

What Do the Cases Tell Us?

- Over time people's interests and goals change.
- Such changes affect each member of a partnership.
- Partnership agreements should reflect these possibilities.
- Unless partnerships are of the limited kind, every partner is fully liable for the other partners' actions.
- Experience may make a person more wary and conservative.
- Corporations provide a vehicle to permit shared ownership.
- Corporations limit *personal liability*.
- The corporate business organization allows a way for ownership to change without disrupting the business.

CHANGE IN A BUSINESS'S STRUCTURE

Why Change a Business Structure or Organizational Pattern?

Among the reasons to change the organization of a business are the following:
- Finances.
- *Perpetuation* of the business.

- *Limitation of personal liability.*
- To change control of the business.
- To divide the assets of the business.
- To bring additional management expertise to the business.
- To permit an owner to spend less time operating the business.
- To provide *employee incentives* and rewards.

Financial Considerations for Changing the Form of a Business

Changing a single proprietorship to a partnership or to a corporation could be done to bring additional finances into the business. This can be done through additional stock sold in the new corporation, or in the case of a partnership, doubling the capital by an equal partnership investment.

Again, changing the *business structure* might permit the removal of personal capital from the business without crippling it. This could be through substituting new capital for the old. In the case of divided assets, the capital was given to the children rather than removed from the business.

A business that was not capable of incorporating in the beginning could have financial success which would allow the original desire to incorporate to become possible. The corporate form would then permit shared ownership, shared management decision making, and limitation of liability.

Perpetuation of the Business. The continued life of a business is easier under incorporation than under sole proprietorship or partnership forms of organization. The corporation continues as an entity even as ownership changes through sale of stock or inheritance. In either of the other business forms, a death can cause an unwanted sale to settle *estates* or to pay inheritance taxes.

Disagreement between partners can break up a profitable business. When the problems no longer need to be solved nor demand the partners' time and energy, personal problems can start to appear.

Limitation of Personal Liability. Limitation of liability is a natural change desired by some people which necessitates a change in the business form. The case discussed at the start of this chapter illustrates this point. A partnership of the ordinary type is a full liability on all the partners regardless of which partner incurred the liability. Needless to say, a sole proprietorship is also fully liable for any business action. The corporation form of business limits liability to the money invested. The sale of one's shares of stock transfers that liability to the purchaser.

Changing the Control of the Business. A change of control in the business results from a change in business form. A change in form from sole proprietorship to a partnership divides the control by the number of partners involved. A change from a sole proprietorship to a corporation or from a partnership to a corporation changes control to a board of directors and officers. These are in turn controlled by the vote of the stockholders.

Dividing the Assets of the Business. A change in the business form or structure may be necessary in order to divide assets. For example, if a partner dies a change may be forced in order to settle the estate of the deceased partner. When no prior arrangements for such an occurrence are in place the results can be devastating to a successful business. The sudden demand on the business for half of its capital would put many businesses in jeopardy.

The way Howard came to be a partner in Exeter Grain is an example. Howard started the Exeter Lumber Company with his good friend Michael who was an expert carpenter and business manager. Howard and Michael started the business as a corporation, owning all the stock issued. After seven profitable years, Michael died and Howard was left to operate the business. He had helped at busy times but did not want to operate the business on a full-time basis. Accordingly he sold the stock in the company and purchased a share in Exeter Motors, again leaving the active management to others. As you will recall Exeter Feed and Grain was a spin-off of Exeter Motors. Howard preferred the management of his five farms to that of day-to-day business management. Sometimes the changes in an industry will cause a change in business structure. In Figure 30–1, the farm was too small.

Figure 30–1   The changing structure of the farming industry left this parcel too small for efficient operation.

Bringing Additional Management Expertise to the Business.   The business structure may be changed to bring additional management to the business. Again, Exeter Feed and Grain can serve as an example. When Ted and Howard started the business they had offered Phil the job as manager; however, Phil wanted an opportunity to own a part of the business. Ted and Howard then offered Phil a full partnership in order to gain his services as manager.

Large corporations offer stock options as inducements for hiring their choice as chief executive officer. The recent past is full of such stories.

Permitting an Owner to Spend Less Time Operating the Business.   The structure of a business may be changed in order to allow an owner to slow down. The formation of a corporation would provide the vehicle for a change in management. A partnership, properly constituted, could also provide for the desired change in status. Of course, a sole proprietor could hire or promote an employee to the position of manager with full operating authority. This could be accomplished without a change in structure.

Providing Employee Incentives and Rewards.   A change in structure completed to reward employees or to provide work *incentives* for greater accomplishments has been used to good advantage by many businesses in the past. A well-known cosmetics company and another well-known products distribution company could be cited. The larger a person's sales and the greater the sales of their recruits, the higher their incomes.

It is reported that when Communist countries allowed workers to have a small plot to produce under the private enterprise system, these small plots quickly became more productive than the larger farm areas.

A change in structure to permit employee ownership has had positive results in many American businesses. Reward based on accomplishment promotes greater accomplishment.

How Do We Accomplish Desirable Changes in Structure?

Some moves such as changing from a sole proprietorship to a partnership are relatively simple. Moving from a sole owner to a partnership is accomplished by selling the appropriate share of the business to a purchaser(s) and setting the terms of the partnership agreements. In turn, changing from a partnership to sole proprietorship can occur by one partner buying the other(s) out. Your business lawyer can safely advise you on the legal steps to take in your state.

If the change is to be from a sole proprietor or partnership to a corporation, a lawyer familiar with such matters should be retained to accomplish the necessary steps.

How Can We Structure the Business to Avoid the Trauma of an Unwanted Change?

A great deal of foresight is necessary to avoid all the pitfalls that a young enterprise faces. The chapter on managing business risk emphasizes a number of key concepts.

In particular, corporate life insurance can be paid on the lives of key persons. The partners in a partnership may pay for life insurance on each partner. Proceeds are payable to the estate of the partner as payment in full for that partner's share of the business.

One of the most important documents may be the partnership agreement that sets out the terms of the partnership including the terms in case of buyout or death of a partner. In the case of the change in partners discussed at the beginning of the chapter, no such agreement was in place.

The change when Howard sold his share of the business was swift and dramatic. The loss of unlimited liability against Howard's assets made Phil's cash flow management critical. Cecil, the banker, restricted *operating loans* immediately.

Other items that can make a well thought out partnership agreement very important are such things as:

- sudden economic changes
- sudden health changes of a partner
- lawsuits over some aspect of the business or of one partner's actions

All partnerships should begin with a complete understanding and a workable partnership agreement as recommended in an earlier chapter.

SUMMARY

A business may reach a state of maturity that recommends a change in its structure in order to function properly. Many adverse events can impact on the business and therefore a current agreement is vital.

Many risks can be managed by the purchase of insurance, specifically insurance on the lives of partners, which is structured to leave the business unscathed financially.

SELF-EVALUATION

Change in Structure Quiz

True-False: On a separate sheet, write T for true statements and F for false statements.

1. Corporate life insurance on key personnel is one way to protect a partnership in case of a death.
2. Partnership agreements limit the partners' liabilities.
3. Changes in the business structure should be directed by the company lawyer.
4. Stock options are used by many companies as an incentive.
5. A change in structure may be made to allow an owner to spend less time in the business.
6. A change in structure is necessary to bring in new management.
7. Changing the structure cannot change an owner's liability.
8. A corporate structure contributes to business continuity.
9. The most common reasons for structural changes are financial.
10. Family corporations provide an excellent way of bringing sons and daughters into the business in a way that allows their growth in ownership.

STUDENT ACTIVITIES

1. Write a partnership agreement for a given partnership.
2. Check out rates for term insurance for corporate life insurance.

CHAPTER 31

Should I Sell the Business?

OBJECTIVE

To explore the considerations in selling a business.

COMPETENCIES TO BE DEVELOPED

After completing this chapter you will be able to:
1. Identify the assets of a business that can be sold.
2. Describe ways material and goodwill assets can be valued in a sale situation.
3. Identify favorable times for the sale of a business.
4. List the advantages of a brokered sale.

TERMS TO KNOW

Accounts receivable	Broker
Business's personal value	Cerebral palsy
Closely held stock	Computer locating service
Consummating	Covenant not to compete
Goodwill	Heirs
Intangible value	Invoice
Market value	Maxim
Par value	Personal worth
Stock shares	Tangible value

INTRODUCTION

At some time in the life of every business the owner(s) will have to decide whether to sell. If you decide that the *business's personal value* is higher than the *market value*, you will keep it, but if the market value is higher, you will sell it in order to continue to use what you now have to obtain what you want most.

If you now are ready to consider the question, or anticipate a time when you may consider it, there are some common sense things to look at as you make that important decision. Among the concerns you will consider are the reasons for selling, what other activities you want to engage in, how the business is doing financially, when is the best time to sell, and how to manage the sale.

CASE STUDIES

The Case of the Dealing Dealer

Chuck was a respected local dealer for a major auto company and had built up a successful partnership over 20 years. Suddenly his partner died and the heirs wanted to sell. Chuck did not see how he could raise the capital needed to buy and still operate the business. Knowing the time it would take to settle his partner's estate, he planned a sale strategy.

In order to increase his volume of business, improve profitability in the short run, and reduce the inventory of used cars, he changed his method of sales. From a policy of making good deals and retailing his used cars, he switched to a policy of price over *invoice* on the new vehicle and a wholesale auction price on trades. This way he did not have to build the used car inventory, which would make the sale of the dealership more difficult.

Chuck was honest with customers and dealt openly with the figures. As a result his quarterly sales zoomed and his allotment of cars was insufficient to meet sales. He bought from other dealers through the *computer locating service* and the dealer books were attractive to buyers. Chuck sold favorably and turned a forced sale to his own advantage. Of course, his partner's *heirs* also benefited from this strategy.

Phil Sells Exeter Feed and Grain

Phil's second daughter was diagnosed in July as having a mild case of *cerebral palsy*. In looking for ways to help his daughter, Phil discovered that the state university (over 350 miles away) had the top expert in the country who specialized in these problems.

Phil immediately discussed the situation with his old and new partners, and agreed on a sale price based on the net worth of the business. The sale would take effect when Phil was ready to relocate near the state university. Phil had a lifetime teaching license. He quickly obtained a teaching job close to the state university and returned to his first profession.

What the Cases Tell Us

- Businesses are sold for many reasons.
- Forced sales are not desirable.
- With sufficient lead time a business can be prepared for a change in ownership.
- A sale based on net worth does not reflect the real worth.
- Partnerships should insure their membership to avoid forced sales.
- Partnership agreements should be written to reflect possible sale conditions.

SELLING THE BUSINESS

Although some entrepreneurs buy and sell businesses for a profit, most sales occur for personal reasons. As noted in the case studies, death or health reasons may cause an unwanted sale. In other cases the owner may be tired of the business or may lack the finances necessary to keep the business going. Sometimes a partial sale is made to obtain funds. In your case, you may sell a small business you start in order to go to college. For whatever reason a business is sold, there are many concerns to face. Most sales affect us in some manner, Figure 31-1.

What is There in Your Business to Sell?

Most businesses have value. This value comes from physical property, inventory, *goodwill,* a list of customers, a product patent or copyright, a service style, a franchise, or some other *tangible* or *intangible value.*

Figure 31-1 Selling the family farm is a traumatic experience. Retirement is the reason for this sale.

Physical property is most easily identified and assigned a value. This property can be the business's real estate, tools, equipment, fixtures, and inventory on hand.

Goodwill and lists of customers are harder to value. This can best be illustrated by the author's purchase of an existing insurance agency some years ago. The purchase agreement consisted of the following assets:

- All current records pertaining to said general insurance agency business, all of the daily reports thereof, all of the renewals thereof, all of the policies in force, all of the supporting documents, papers, files, and all other items pertaining thereto
- The furniture, supplies, and equipment now in use in said agency
- The goodwill of said agency, including, but not limited to, insurance expirations and the trade name or names used by Seller in connection with said agency
- The *accounts receivable*

A price on each of the first three items and a *covenant not to compete* provided the total purchase price. The market price of the furniture and equipment would be most easily derived.

To illustrate this point, if the list of business records in the first item would yield net income of $40,000 at the end of the year, the present value of that $40,000 would be about $33,326 assuming today's interest rate of 10%. Negotiation of the sale would start from the $33,326 rather than the $40,000. Presumably some portion of that $33,326 would be a value placed on the records and goodwill.

In cases of a service business, the buyer may have furniture and office equipment and only be interested in the records and goodwill from the purchased business. In either case, goodwill is an estimation on the part of the purchaser as to the potential income from the purchased business. Scott's goodwill for his birdhouse business would be difficult to measure, Figure 31–2.

Inventory or undesired items may be auctioned to determine their current market value or to dispose of assets not wanted by the buyer. For example, a purchaser bought out a business to obtain the location to expand his own business and to provide extra parking. In this case the new owner had no use for the equipment owned by the seller or his goodwill either. An auction quickly cleared the way for the remodeling and new use.

When to Sell the Business

There is a *maxim* that says you should never sell or buy when you have to sell or buy. In other words, forced sales may not be good sales. One should sell at the most opportune time. It is more appropriate to sell the business when it is in its busy rather than its slack season. Psychologically it provides more incentive to the buyer who sees the business's traffic.

A decision to sell the business should be made while the seller is enthusiastic, not discouraged, about its prospects. If the seller is discouraged about the business, it will be undervalued and the sale may not receive its best representation.

On the other hand, the owner selling during the busy season will need to be careful not to get so involved in the day-to-day operation that he or she fails to put forth the effort to complete the sale. The daily routine can be shifted to another employee and extra effort channeled to showcasing of the business and providing information and insights to the prospective buyers.

If the sale occurs during the slower season, showcasing the monthly business reports for the busy season will serve to offset the seasonal lows actively observed.

Figure 31-2 Scott assembles a birdhouse for sale. (Courtesy Joyce A. Niles)

How to Sell the Business

There are three common methods for selling any business. Each of these has its own advantages and disadvantages. The three methods are *broker* sale, sale by owner, and auction.

Sale by a Broker. A real estate broker who specializes in selling businesses may be the best person to dispose of a business and its assets. In this type of sale, the broker solicits prospects and shows the business. He or she also advises the owner on how to prepare for the sale, how to finance it if required, and other matters concerning the sale. For these efforts the broker receives a percentage of the sale price. This percentage may vary or may be calculated on a sliding scale. The percentage is from 5–7% in most cases.

The advantages of a sale by broker are:

- advertising by the broker.
- showing by the broker.

- financial advice from the broker.
- leaves you free to conduct your business.

The disadvantages of a brokered sale are:

- cost.
- business interruptions due to prospective buyers.
- buyers inspect your books, which you may not wish to disclose.

Sale by the Owner.   When the owner sells the business, he or she controls all aspects of the process. The services of a consulting business may be used to obtain marketing advice.

The advantages of sale by owner are:

- lower cost.
- total control of process.

The disadvantages of sale by owner are:

- lack of professionalism in process.
- time required away from developing business.
- need to plan advertising.
- time needed to show the business.
- need to enter financing efforts.

Sale by Auction.   Sale by auction is a quick way to determine market value. It provides open bidding to establish the price someone is willing to pay on that day. The auction may provide inadequate time for a prospective buyer to establish needed financing, but it does provide a selling price and a change in ownership.

The advantages of an auction are:

- quick solution to the sale problem.
- time saved in effecting the sale.
- competitive bidding.

The disadvantages of an auction are:

- cost.
- lack of control of selling price.
- number of nonbuyers.
- total disruption of the business on the sale date.

Making a Partial Sale

Sometimes a sole proprietorship decides that a partial sale is the best option for a business. In such cases there are two common alternatives, the sale of a partnership interest or the sale of stock through an incorporation.

The Sale of a Partnership.   The partial sale of a business through partnership provides the opportunity for the owner to double the capital of the business. It also provides the opportunity to double the management skills of the owners. The sale divides the responsibility for operation as well as dividing the workload. The partial sale, however, slows down the decision-making process since more people are involved. Remember if you are the boss you get to turn out the lights, Figure 31-3!

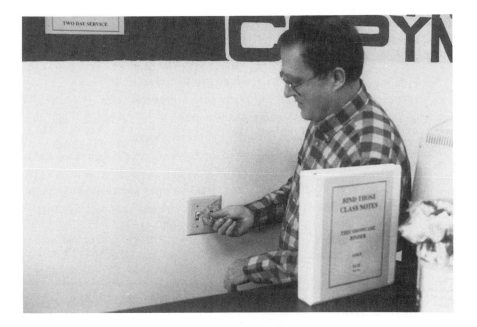

Figure 31-3 If you are the boss, you get to turn off the lights. (Courtesy Don Steele)

We urge any business owner forming a partnership to consider a written partnership agreement spelling out the procedure to be followed in ending the partnership. Some of the risks can be shifted to insurance. Many business partnerships pay life insurance premiums on the lives of the partners, with the proceeds going to the partner's heirs as complete compensation for their rights to partnership assets. The surviving partner(s) then are free to continue the business without fear of a forced sale.

A Sale of Stock in the Business.   The second method of *consummating* a partial sale is by forming a corporation and selling stock. Public corporations must be a minimum size to be eligible for open trading on the stock exchange. However, a modest sized corporation can be formed—in the authors' state the assets must total at least a thousand dollars.

Stock shares are issued at a *par value,* which may or may not be the market price. As owners of shares, consider our original question. Is *personal worth* more than market value? If personal worth is less than market value, then a sale can occur. *Closely held stock* is found in a majority of smaller corporations and is infrequently sold or marketed.

In the case of a corporation, the decision-making system is slowed as the stockholders vote on policy matters and elect and instruct executive officers. Responsibility is divided as each stock owner is liable only to the value of stock owned.

SELF-EVALUATION

Sell the Business Quiz

True-False: On a separate sheet, write T for true statements and F for false statements.

1. One advantage of an auction is the competitive bidding.
2. Another advantage of the auction is the price control it gives.
3. A partial sale sometimes may be a solution to forced sale.
4. An auction of a business takes more time than a brokered sale.
5. Selling the business yourself requires you to manage the advertising and showing.
6. Timing the sale has very little effect on the potential price.
7. A sale when the owner is enthusiastic will be more successful than when the owner is discouraged.
8. A disadvantage of a brokered sale is the time you must spend showing the business.
9. An advantage of a brokered sale is that you can continue business as usual.
10. Sale by the owner will cost less but be less professional.

STUDENT ACTIVITIES

1. Interview a real estate broker who handles business sales and ask how businesses are marketed.
2. Collect the real estate ads from your area newspapers and analyze the types of businesses that are for sale.
3. Identify the reasons for sale in as many cases as you can.

CHAPTER 32

A Last Look at Entrepreneurship

OBJECTIVE

To review the major concerns for a beginning entrepreneur.

COMPETENCIES TO BE DEVELOPED

After completing this chapter you will be able to identify the major strengths of the course material presented and make suggestions for changes.

TERMS TO KNOW

Since this is a summary chapter, no new terms are presented. Instead, if you are unsure of meanings, please refer to the glossary that follows the next chapter.

INTRODUCTION

In this summary, we want to present a review of the major points one needs to consider in deciding to start one's own business. These will be summary statements only. If you wish to review a particular topic in greater depth, many references are available. Your teacher should have information about additional resources on the topic of entrepreneurship.

The American dream is to become self-sufficient and entrepreneurship gives us all a place to start. Many entrepreneurs get their start by expanding a hobby, Figure 32–1.

Since this is a summary chapter, no new case studies will be presented. Case studies were included to provide an interesting way of bringing out key points important to the chapter's content. When this author visited the field test site, Donald Connelly, the teacher, said, ''Students, meet Phil.'' Immediately the group said Phil always seemed to have a lot of trouble. I assured them that Phil had many good times but that discussing his mistakes led to the key points faster.

Figure 32–1   Crafts have provided the entry point to entrepreneurship for many people. Sue works on one of her baskets. (Courtesy Susan M. Peters)

THE PERSONAL SATISFACTION IN SELF-EMPLOYMENT

Many people express a desire to start their own business so they can be their own boss. Why not? It is the American dream! However, because of fear, they lack the spark that gets them to risk it all in order to begin. (This may be the entrepreneurial attitude.) Entrepreneurship comes about when the individual dares to take the chance and dares to fail (reread Figure 1–3). The entrepreneur asks questions to find out what is happening and applies creative thinking in getting the job done!

We hope the use of this text helps you look accurately at the advantages and disadvantages of self-employment. More important, we hope to have provided you with ideas to use as you plan what could multiply your chances of success.

Self-employment, full-time or part-time, gives you the chance to be creative and try out your superior idea or better way of doing things. It is estimated that 80% of new jobs created in America are provided by small businesses, but almost every week we see the announcement of a factory shutdown and the loss of local jobs.

The key components of working for yourself are:

- You control the enterprise.
- You set the level of competition.
- You set the hours you work.

- You reap the profit (or loss).
- You can develop your good ideas and be creative.
- You can call on experts of your choice.
- Your decisions are the ones that affect your life.
- You set your own policies.
- You get to turn out the lights.

Disadvantages you should be aware of:

- You provide all the finances.
- Changing your mind will be costly.
- Most businesses lose money in the beginning.
- Many other people may have the same ideas (competition).
- You are accountable to everybody: your customers, lenders, employees, and your family.
- You, not the lender, take the risks!

WRITING A BUSINESS PLAN

Starting a sole proprietorship is a relatively simple matter, but a good start requires careful planning. In this section we want to point out the major headings of the business plan and mention the kinds of information you need to provide under each.

By this time you should have a well outlined business plan. It is your road map and permits meaningful communication with a lender. Let's take a look to see if your plan includes each of the items shown in Figure 32-2.

GOAL SETTING

We should always remember the old proverb—every journey starts with a single step. But first we need to know where we are going. We started an enterprise in the first place because we had goals—goals that may in fact have been fuzzy, like ''I want to make a million dollars and retire at the age of 40.'' Let's get rid of the fuzzy ideas and set short-term, or immediate, and intermediate goals so that we can reach the long-term goal. Each of these goals must focus on a part of the plan. Writing the business plan is the confirmation of the goals you have set for yourself. One young man's idea is shown in Figure 32-3.

We must remember that as we set these short- and intermediate-term goals, we must make them as simple as ABC.

A—*Attainable*. We may get discouraged if we don't achieve some measure of accomplishment in the short run.

B—*Believable*. We must believe the goal is worth achieving and that we can in fact achieve it.

C—*Challenging*. Without making ourselves stretch a little, we will dare too little and that will slow our progress.

As you start to review your goal setting, remember to look at your unstructured time as shown in Figure 32-4. Your free time is time you are not using for sleep, meals, school activities, and study. The illustration is merely an approximation.

DEVELOPING THE GREAT IDEA

Organizing my business venture
Setting my business goals
Recognizing my resources
Identifying the competition
Finding my market
Promoting my business
Accounting for my business
   (my financial plans)
Managing the details
   My legal concerns
   My taxes
   My location
   My facilities
   My cost and pricing
   My scheduling
   My business procedures
My business relationships
When I hire others
Do I use computers?
My business grows
Business continuity plans

Figure 32–2   My business plan outline. (Courtesy The Three Entrepreneurs)

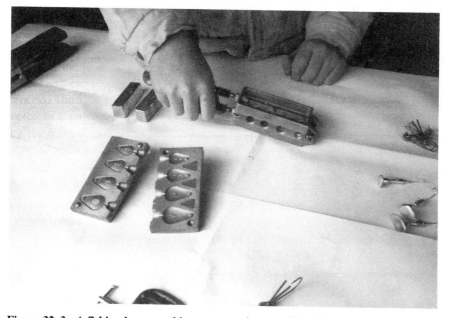

Figure 32–3   A fishing lure was this young man's great idea. (Courtesy Janet Rausch)

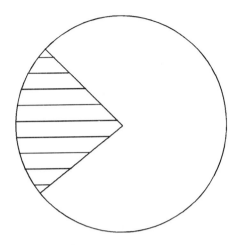

Figure 32–4 My free time. (Courtesy The Three Entrepreneurs)

When you have inventoried your weekday and weekend uses of time, you must set priorities. List your top 10 and start working towards them daily. Today's top 10 goals are illustrated in Figure 32–5.

Figure 32–5 My top ten goals. (Courtesy The Three Entrepreneurs)

Remember, as you complete one of today's priorities, cross it off and add a new tenth item. Items not completed today either disappear or move to tomorrow's list if they are still on the top 10.

ORGANIZING OUR BUSINESS

Businesses are classified in several different ways. First, we can classify them by what they do—manufacturing or production, sales, or service. A second classification is based on the type of ownership: sole proprietorship (owned by a single person); partnership (owned by two or more individuals); and stock companies (owned by shareholders).

The sole proprietor has full management responsibility and decision-making autonomy. This person also has to furnish all of the capital and in turn gets all the profit or suffers the losses should they occur. It is the simplest form of business organization.

The partnership is a case of divided assets, divided decision making, and divided profit or loss. Unless it is a limited partnership, each of the partners is liable for the others' actions. Partners can be active (operating), or silent (having only financial interest) in the management of the business.

The stock company is operated by a board of directors and decision making is more involved. Liability of stockholders is limited to the extent of the value of shares held. This is the most complex form of business organization.

OBTAINING A BUSINESS

You can purchase a business and continue operations with immediate cash flow and returns or you can start a business from scratch and mold it in your own direction. You will need to examine these concepts in choosing the route to business ownership.

A purchased business has:

- an existing cash flow.
- name recognition.
- reputation and goodwill.
- established facilities.
- established suppliers.
- an experienced staff.
- eliminated one competitor.
- loyal customers.

When you build a business from a new start, you:

- have no bad history to overcome.
- have the opportunity to be fully creative.
- get to choose the location.
- organize the facilities.
- determine the basis for entry into a competitive market.
- hire and train employees.

Mind, money, and muscle represent our personal resources in starting an enterprise. As illustrated in Figure 32-6, mind represents creative ideas, plans, and management; muscle represents the time, effort and skills you contribute to a successful business; while money represents the fiscal or physical resources committed to the enterprise. One needs to inventory available resources and then consider which can be obtained in order for the business to flourish. You probably would not want to start a business that requires more time than you have available to devote to it.

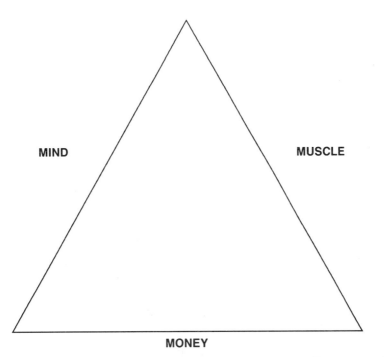

Figure 32–6 Entrepreneurial resources. (Courtesy The Three Entrepreneurs)

MARKET RESEARCH

Before starting a business, think about the market for the product or service. The facts in Figure 32–7 should help to estimate the chance for success of a given type of business in a small city (8–10,000 population). These figures were taken from census data. The figures indicate the average population for the listed types of businesses in the United States. Figures are in thousands.

BUSINESS	POPULATION IN THOUSANDS
Auto & home supply	6
Automatic merchandising machine operators	43
Automobile dealers—new and used	9
Automobile dealers—used only	20
Bakeries	13
Boat dealers	57
Books	25
Building materials	7
Camera, photographic supply	58
Candy, nut, confectionary	45
Children/infant wear	44

BUSINESS	POPULATION IN THOUSANDS
Dairy products	46
Department	23
Drugs	5
Family clothing	13
Floor covering	21
Florists	11
Fruits and vegetables	78
Furniture	8
Gasoline stations	2
Gifts, novelties, souvenirs	10
Grocery	2
Hardware	12
Hobbies, toys, games	30
Home furnishings	17
Household appliances	22
Jewelry	10
Luggage, leather goods	122
Mail order	31
Meat and fish	21
Men and boys	13
Miscellaneous general merchandise	18
Mobile home dealers	49
Motorcycle dealers	50
News dealers and stands	116
Nursery—lawn and garden	30
Optical goods	22
Radio, TV, and music	8
Restaurants	2
Sewing, needlework, and piece goods	6
Shoes	6
Sporting goods, bicycles	12
Stationery	48
Used merchandise	13
Women's clothing	5
Women's ready to wear	5

*Statistical abstract of the United States: 1985

Figure 32–7 Average population in thousands for selected businesses in in the United States. (Courtesy The Three Entrepreneurs)

Market research goes further than determining how many gas stations are in a city with a population of 20,000. Market research looks at the purchasing habits of the target population, where most purchases are made, their quality, style, and other characteristics. Most new businesses need to decide how they wish to compete on quality, price, style, or service. In effect,

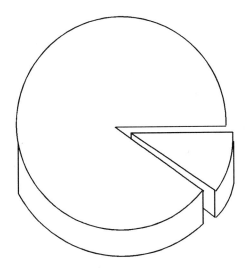

Figure 32–8 My piece of the cake. (Courtesy The Three Entrepreneurs)

market research helps define your market niche. Figure 32–8 illustrates the idea that each business in a field has a slice of the total cake.

DECISION MAKING

An entrepreneur has to make a lot of decisions in starting a new business. Often this is done with little experience or few records to guide the entrepreneur. Family and friends are essential support groups during this trial period. Each section of the business plan outlined in Figure 32–2 calls for decisions. How well these are planned and implemented may spell success or failure for a new enterprise.

ADVERTISING

Deciding what an advertising and promotional budget should be when resources are scarce is tough but important. Goods and services are scarce commodities and economics is the allocation of resources. Advertising is presented to convince others to buy our product or service instead of the competition's. Many methods and media are available for promoting the enterprise. Experience may be needed to assess the media that is most effective.

Newspaper coupons are one method of assessing how many customers reacted to an advertisement. An offer made only on the radio could measure a response to that media and a different television promotion could measure that media's impact on the business. The old adage, ''Advertising doesn't cost, it pays,'' has a lot of supporters.

COMPETITION

Competition will be examined in an adequate market study. However, competition should also be considered directly as a question of supply and demand. It is obvious that a new business in an established line must get customers from other businesses or convince customers who did not use the product before to make the purchase.

ACCOUNTING PROCEDURES

In starting a new enterprise on a small scale, either a single entry bookkeeping system (obtained from an office supply store) or the more detailed double entry system could be put in place. The decision should be based on the volume of business and the reporting system needed. Double entry provides a more adequate base for reporting and planning information.

Records are essential for income tax purposes and ownership reports, and are a basis for future projections. Everyone in the business must know how to record transactions and prepare documentation.

An early decision on credit policy for your business is essential; accounts receivable are a constant worry when dealing with cash flow problems. Most businesses should not be in the lending business. Credit can increase sales but requires thoughtful management.

Adequate records help in decision making. At your present stage in entrepreneurial activities, you probably have very little experience on which to base future operational decisions. If you do have some records, even limited ones, they will be helpful in projecting expanded operations because actual costs are superior to educated estimates.

RULES, REGULATIONS, AND TAXES

Every business person must observe governmental regulations and meet tax obligations on time. In planning your business operation you need to become informed of local ordinances, and state and federal laws that apply to your intended business operation. In many cases, a company attorney is a wise investment.

COSTS AND PRICES

A common area of concern and error in a new business is that of accurately determining costs and prices. Many young or inexperienced entrepreneurs underestimate the cost of doing business and in the process underprice their product or service. This threatens their business success.

In the recent past, price increases have been accepted with little question. In today's economy it would be harder to raise prices with immunity. Wise managers would probably want to overstate the price slightly in the beginning to prevent an increase when the cost-price relationships are determined.

SCHEDULING BUSINESS OPERATIONS

Scheduling can have a great effect on cash flow. In fact, in the beginning a business has a negative cash flow. The cash flow position may also affect personnel management decisions. For example, products that take considerable time to assemble may require advance estimation of sales and, therefore, productive lead time. A simpler product could catch up with sales through a judicious use of overtime or addition of a second shift. In the case of service activities, close coordination with sales is needed to pace the timing of the service in order to keep customers pleased with your operation.

The second aspect of scheduling has to do with controlling costs by limiting inventories of raw materials and finished goods. Scheduling deliveries to provide only a comfortable margin between receipt of raw materials and shipment of finished goods helps promote cash flow.

In a service business, such as installing kitchen cabinets, this would mean receiving the cabinets the day before the business was to install them and contracting payments on completion.

Figure 32-9 again illustrates the cash flow cycle. In the first quarter the cash is flowing out for raw materials and production. As goods are finished and sales are made, cash flows back into the business. Ronnie approaches the cash back position, Figure 32-10.

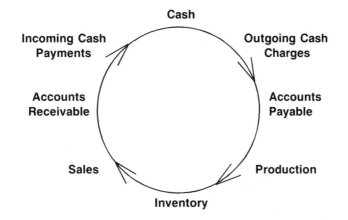

Figure 32-9 The cash flow cycle. (Courtesy The Three Entrepreneurs)

BUSINESS PROCEDURES

A sound set of business procedures will provide consistency and give customers confidence in the way the business operates. The same consistency is important in working with employees. They need to know there is consistent treatment under clearly stated policies. When your business grows to the point of adding employees, a policy manual is needed to guide growth.

HUMAN RELATIONS

Business needs good human relations in all areas including the way it treats its employees, customers, and prospective customers. Consistently fair policies and consideration for the dignity of the individual are always appropriate. Long-term survival of a business is more likely when those who operate the business maintain a high moral level.

Figure 32–10 Ronnie puts the finishing polish on a truck
wheel. (Courtesy Ronnie Orem)

BUSINESS RISK

When we enter a business we assume different kinds of risk. Risk management suggests that we control some risks through risk reduction, risk transference, or risk avoidance. We may have to retain some of it as best we can.

Risk avoidance means that we avoid some risks by avoiding their cause. Bad debt risk can be avoided by doing a strictly cash business. We need to realize, however, that we may do less business than if we extended credit.

Risk reduction means that we reduce the risk of loss through formal actions such as safety training, machine guarding, or similar activities.

Risk transferring is paying a premium to an insurance company which will assume the risks for you. Examples of pure risk that a business may choose to insure against are fire, theft, hail, windstorm, vandalism, and even the life of the entrepreneurs.

Risk retention refers to uninsurable risks or risks the entrepreneur may decide to retain rather than pay a premium to cover.

COMPUTER USE

When to computerize the business is an important management decision. Computers are an appropriate tool to employ when:

- adequate software to operate your record-keeping system is available.
- many repetitive tasks are involved.
- the computer will lighten the task, speed up operations, or make forecasts and projections in an economic manner.
- you or one of your employees feels comfortable when using their capabilities.

CHANGES IN THE BUSINESS

Business growth, a change in your interest, or the need for more capital may all be change agents for your business. Changes may take place in the business organization or structure or through the sale of the business itself. You may decide to sell in order to attend the university or you may have someone else manage the business while you are in school.

Many times the need for further capital may produce a change in organization through the addition of a partner or the formation of a public corporation.

IS ENTREPRENEURSHIP FOR ME?

The next chapter looks at several measures of interest in entrepreneurial activity. As you study this chapter remember:

- Persistence pays off. Extra effort often makes the difference between success and failure.
- If you do not proceed with an entrepreneurial enterprise, you are not flawed and have not wasted the time you have spent on this course.
- *Nothing in the world will take the place of persistence.*
  Talent will not—nothing is more common than an unsuccessful man with talent.
  Genius will not—unrewarded genius is almost a proverb.
  Education will not—the world is full of educated derelicts.
  Persistence and determination alone are omnipotent.

Author unknown

Finally, remember to use your resources and entrepreneurial recipes to gather rewards. Rewards may be profit and satisfaction. To reach the final rewards you must pay attention to the details and keep your priorities in order. Your family will be growing while your business is growing. Don't neglect either one!

If you never dare to dream there is no way you can make your dreams come true. Good luck as you promote your better idea.

TEXT EVALUATION

Please rate each chapter as it ranks in helping you understand and develop entrepreneurship concepts. Excellent = highest, Good = helpful, and Poor = needs improvement.

Chapter	Subject	Excellent	Good	Poor
1	Self-employment			
2	How to Start			
3	Goal Setting			
4	Business Organization			
5	Buy or Build			
6	Three Resources			
7	Competition			
8	Market Research			
9	Promotion			
10	Accounting			
11	Single Entry			
12	Double Entry			
13	Start-Up			
14	Records and Projections			
15	Credit			
16	Legal Concerns			
17	Taxes			
18	Location			
19	Facilities			
20	Costs			
21	Prices			
22	Scheduling			
23	Procedures			
24	Customers			
25	Personal Contacts			
26	Hiring Others			
27	Risk			
28	Computers			
29	Growth			
30	Business Structure			
31	Sell the Business			
32	A Last Look			
33	Am I Suited?			

CHAPTER 33

Am I Suited For Entrepreneurship?

OBJECTIVE

To assess your entrepreneurial interest.

COMPETENCIES TO DEVELOP

After completing this chapter you will be able to:

1. Assess your interest in entrepreneurship.
2. Identify at least 10 characteristics of successful entrepreneurs.
3. Identify eight skills needed by entrepreneurs.

TERMS TO KNOW

Assessed
Conservative
Established
Endeavor
Lease
Spin-off
Support Group

INTRODUCTION

Self study is a means of evaluating your entrepreneurial potential. In this chapter we list a series of questions that will provide such a self study.

CASE STUDIES

Phil Takes a Job

We met Phil several times in the previous chapters. He accepted a partnership in the Exeter Feed and Grain business, a spin-off of Exeter Motors. What were the factors that led Ted and Howard to offer Phil a full partnership in the new business?

- Phil was a successful teacher in the local school.
- He was well-respected by the county farmers.
- He had an evening school program with over 80 farmers in attendance.
- Three instructors working with 75 veterans were under his supervision.
- He was married and active in community affairs.
- He had taught in two other schools.
- He was a veteran with noncommissioned officer experience.
- He was interested in staying in the community.
- They knew his brother who lived in a nearby community.

Now, why did Phil take the partnership?

- If he ran the place, he would share in the profit (*profit,* not loss).
- He believed in himself.
- He believed he could work effectively with the farmers in the community.
- He was well acquainted with the trade territory from working with adults and veterans.
- He was acquainted with the livestock problems in the area.
- He was acquainted with the franchise feed company. They sold the top product and provided educational support.
- He had a strong livestock program.
- He had a B.S. degree from the state university plus some graduate work.
- He had had experience with the grain business.
- He was ready for a change or challenge.

Kermit Decides on Sweet Corn

Remember Kermit from Chapter 1? If not, reread his case study. Why did Kermit reinvest the money earned from his sow and litter?

- Kermit was ambitious. His goal was to get a college education, and his grandmother's income would not cover the cost. His grades wouldn't earn him a scholarship, so he decided he could earn the money.
- He saw an opportunity in the vacant acreage nearby.
- He thought he could talk the developer into a rental agreement.
- He had a *support group*—the agricultural education teacher, his grandmother, and his uncle were available.

- These people backed him in tangible ways—influence, advice, equipment, and supervision.
- The risk was manageable—if he couldn't sell, he could feed. Money risk was low. The largest input was muscle—his labor.
- He could *lease* the equipment needed (to be paid for by the results of his labor).
- He had a nearby market.
- He had a detailed plan.
- The project interested him.

What Do the Cases Tell Us?

What do we learn from these case studies? Phil's place in the business when it opened is illustrated in Figure 33-1. Phil provided half of the labor or muscle as the business started with one employee. Phil invested one third of the capital (money) and was the manager (mind); Ted and Howard provided occasional help as their schedules permitted.

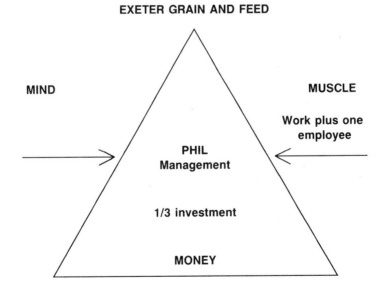

Figure 33-1 Phil's place in the triangle.

- There are many reasons for going into a business venture.
- Goal setting is important. Make the road map, then follow it.
- Planning is important, and should be done realistically and *conservatively.*
- Financing is important. Inadequate funding is a cause of failures. Kermit planned a way to reduce upfront capital needs.
- Good human relations skills are a viable business asset. The ability to get along with other people is important to every *endeavor.*

- A good reputation is also a valuable asset in a new business.
- Risk must be *assessed* and managed at a reasonable level. Phil found the *spin-off* of an *established* business less risky than starting from scratch.
- Self-confidence is an important ingredient in starting an enterprise and should be based on abilities.
- Knowledge in the area of the enterprise selected is essential.
- Successful performance in one endeavor is the best predictor of success in a similar new one.
- Supervisory experience aids in management.
- A knowledge of the market and its area is an important business tool.
- Willingness to learn is essential. Most entrepreneurs have to learn new skills and technical knowledge as they grow into their business.

How would you relate each of the points made in the case studies to the three sides of our entrepreneurial triangle? Muscle is often needed, Figure 33-2.

Figure 33–2 Scott provides a janitorial service. (Courtesy Joyce A. Niles)

WHAT SKILLS DO ENTREPRENEURS NEED?

1. Problem-solving skills include the ability to define the difficulty or problem, the ability to analyze its components, the ability to gather relevant resources, evaluate possible solutions, and choose a plan of action. This ability may also involve identification of experts who can provide information or suggest problem solutions.

2. Human relations abilities are important because every business needs customers. They pay the bills and determine whether or not a business flourishes or flounders. Entrepreneurs need human relations skills, perhaps more than others.

3. Accounting skills are needed by anyone contemplating a business venture. Careful financial planning is a must because most businesses lose money in the first two years. The ability to acquire capital is largely dependent upon the quality of the business plan and the financial forecasts that the plan presents.

4. Decision-making skills are needed by the entrepreneur. If the owner does not make the decision, it doesn't get made.

5. The entrepreneur must have good communication skills to be able to communicate with employees, public, and customers. Without communication no one will know you have a product or service available for purchase.

6. Sales skills are important for selling a product or idea. Without sales, the business will fail. A product or service has to be available and then marketed.

7. Performance skills must be developed within the enterprise unless capital is adequate to fund employees with the skills.

8. The ability to seek needed technological help when necessary is helpful.

9. The ability to persevere in the face of adversity.

10. Management abilities are essential to keep the many different rings of the business circus in operation at the same time. The different parts of the enterprise must be balanced and running smoothly.

AFFECTIVE
Human Relations
Perseverance

COGNITIVE
Technical Knowledge
Decision Making
Communications
Management
Problem Solving

PSYCHOMOTOR
Performance Skills
Sales
Accounting

Figure 33–3 The three domains of business skills and abilities.

Figure 33–3 highlights entrepreneurial skills and their nature. List the skills you will need. Did you include most of these skills? Did you list other skills in addition? Let's try another way to identify the skills needed to operate a business venture.

SELF-EVALUATION

HOW CAN I ASSESS MY POTENTIAL AS AN ENTREPRENEUR?

Which of These Characteristics Do I Possess?

Self-Confidence.   Do I believe I can make or do something better than others? Am I so sure of it that I can risk my comfortable position to take on the world?

Perseverance and Determination.   Am I a person who sticks with a project until it is finished? Do I persist after all others quit the task? If someone says it can't be done, do I do it just to show them they are wrong?

One of the authors and a friend, as college students, removed trees killed by an early winter storm. A 14-inch Chinese elm came down easily and worked up quickly, but the stump was stubborn. The householder suggested we give up and he would get someone else to bulldoze it out later in the spring. We persevered and eventually split and removed the stump. We were tired but triumphant. We had contracted a service and we delivered.

Resourcefulness.   Am I able to solve problems? Do I like the challenge of a new activity? Am I able to tap the expertise of others?

Ability to Take Calculated Risks.   Do I look for risks that can be reduced to a manageable level? Entrepreneurs are risk takers, but with risks they feel they can influence. When the risk can be influenced by confidence and know-how, they are not really considered risks.

Initiative.   Do I see something that needs doing and do it? Am I likely to propose action? Do I like to get the ball rolling? Am I restless while waiting for things to get started?

Flexibility.   Do new ways of doing things appeal to me? Do I like new suggestions on how to do a job?

Vision.   Can I see the whole picture? Can I visualize how all of the parts of an enterprise fit together?

Optimism.   Do I expect things will always work out for the best? Will tomorrow be brighter?

Creativity.   Do I like to do things a new way? Make the item a different way or see a change in the routine?

Independence.   Do I have a high need for recognition or am I comfortable in going it alone? Is the challenge of the task enough motivation for me?

Health.   Do I have lots of energy and good health?

Organization.   Do I keep good notes? Can I find all my possessions when I want them? Is my room orderly? Do I have a system to keep track of things?

Problem Solving.   Do I enjoy the challenge of a problem? Am I able to figure out how to do a job in spite of the road blocks? Am I good at analyzing a situation?

Status or Need for Recognition. Do I sulk if someone else gets the credit? Can I grin and keep going when someone else basks in the spotlight? Is my knowledge of my valuable contribution enough to keep me motivated? Am I still motivated when the job turns out to be hard work? Figure 33–4 illustrates the idea.

Figure 33–4 Remember not all entrepreneurship jobs bring status and recognition. Some are just hard work! (Courtesy Lester Bean)

Emotional Stability. Do I blow my stack easily or am I a cool-headed person? Can I keep the job going when the pessimist is ready to quit?

Commitment. When deadlines come, do I work extra hours without any thought other than getting the job done? Does the enterprise come first or nearly so? Do I believe tough times don't last but tough people do?

Goal Setting. Do I set goals for myself and then work hard to reach them? Do I set deadlines for finishing tasks? Does setting deadlines help me accomplish goals?

These and many other characteristics are present in entrepreneurs. If you answered many of these questions in the negative, then your personality patterns may not match those of most successful entrepreneurs.

GLOSSARY

These definitions apply only to this text and are simplified to suit that purpose. For complete definitions in other contexts, please consult a dictionary. Where two definitions are separated by a "–" the second definition represents the meaning in the second context.

Account sheet—ruled pages for accounting entries

Accountant—a person paid to keep business records and complete financial reports

Accounting expenses—a business expense for keeping business records

Accounting procedures—rules followed by accountants in bookkeeping

Accounting rules—how accountants handle transaction reports

Accounting system—how business records are kept

Accounts—business records

Accounts payable—accounts owed by the business

Accounts receivable—money due the business

Accrual basis—an income tax reporting method based on reporting income at sale and expenses as incurred without regard for payment dates

Acrimony—name calling or other verbal abuse

Adamant—insisting

Adaptations—changes in the normal delivery system

Adept—proficient or skilled

Adjunct—something joined or added to, but not essential to, the business

Administrative expenses—the expenses involved in running a business above the cost of production (such as accounting, and so on.)

Adversity—any difficulty or problem in the way of progress

Advertising—the process of attracting public attention to a product or a business

Advertising agency—a private business helping other businesses conduct advertising.

Advertising campaign or program—a planned advertising schedule

Advertising copy—the material to be used in an advertisement, to be printed in the newspaper or other media, or to be read on the radio

Advertising media—methods of presenting advertising copy

Aesthetics—relating to beauty

Age Discrimination Act—1967 federal law prohibiting age discrimination in employment for persons aged 40–70

Agent—person authorized by a business to make contracts with customers

Agri Data Network—data system that can be accessed by members' computers

Alienate—cause to no longer be friendly

All-risk insurance—covers all normal risks

Alternative—offering a choice

Alternative price—a different price

Anticipated sales—projected sales

Appraiser—a person who estimates current market value

Appreciation—the increase in value of an asset

Aptitude—ability to learn, for example, musical aptitude

Arbitration—disagreement settlement through mediation

Area planning board—local governmental agency that regulates business locations

Arson—fire deliberately set

Assessed—measured or evaluated

Asset—something of value owned

Attitude—positive or negative feelings toward something

Audit—checking the accounting accuracy of others

Automobile row—several competitors located on the same street or highway

Avert—avoid

Backup—an alternate system that one can call on if the main system fails

Bad debt—when an unpaid bill is judged uncollectable

Bankrupt—unable to pay debts because liabilities exceed assets

Balance sheet—a financial statement showing assets and liabilities

Basic coverage—the coverage usually required, such as auto insurance for property damage

Behavior—the actions or reactions of certain persons or things under specified circumstances

Belligerently—combatively or abrasively

Benefit package—the benefits paid to or for employees as part of their compensation

Benefits—compensation in addition to salary or wages

Binding agreement—written contract

Black cloud—item perceived as bad news

Block man—a company representative covering a specific area

Board of directors—a group elected by stockholders to manage a corporation

Bond or bonded—building stronger ties with the business

Bookkeeping—recording an entity's financial transactions and events

Bottom line reasoning—refers to the profit motive

Break—the scales were calibrated by five pound increments, so anything between marks is extra

Break-even analysis—the point where the dollar amount of sales is exactly equal to costs

Broker—Real estate dealer, in this case specializing in business

Budgeting—financial planning in decision making

Budgetary constraints—limitations of planned financial commitment

Building permit—legal permission to build, remodel, or tear down a building

Built-in bias—a selfish reason for acting

Business associates—individuals who are affiliated with a business

Business deduction—a business expense

Business entity—may be owned by an individual, a partnership, or shareholders; its purpose is to earn a profit

Business ethics—business moral practices

Business expansion—business growth

Business image—the concept held by the public, especially customers and business associates, of a business

Business logo—a symbol providing public recognition

Business plan—a written plan recording the decisions one makes in starting a business; a road map

Business policy manual—a written compilation of company policies

Business referral—customers sent to your business by others

Business venture—chancing the start of a business with no certainty of success

Campaign—a series of operations designed to promote a business or its products or services

Cancellation penalty—the cost of canceling a contract before it has expired

Candor—frankness

Canvass—to visit and request business

Capital—the funds contributed to a business by owners or lenders; wealth

Case study—real life stories emphasizing concepts and economic principles

Cash basis—income tax reporting on the basis of cash payments and receipts

Cash cycle—the way cash flows into and out of a business

Cash disbursements—money paid out

Cash flow—a budget that reflects the timing of expenses and income

Cash flow problem—when more money is going out than coming into the business

Cash flow projection—careful estimates of costs and returns by month for the year

Cash flow sheet—a sheet that details cash flow expectations

Cash receipts—money taken into the business

Cash settlement—money paid to settle a claim

Catastrophic loss—a disastrous loss, for example, damage caused by a tornado

Categories—groups or classes of similar items

Cerebral palsy—brain damaged condition

Chamber of Commerce—an organization of local businesses that promotes the community

Chain—a company having outlets in many cities

Chattel—tangible property such as equipment

City zoning ordinance—a law that determines where businesses may be located in the city

Civic activities—actions for the good of the community as a whole

Civil Rights Act of 1964—federal law prohibiting discrimination in employment

Clearance price—reduced prices to clear merchandise

Clientele—customers for your product or service

Clients—customers

Cloudburst—rain of 3 or more inches per hour

Coinsurance—the insured provides some of the coverage

Collection agencies—a private business that collects accounts receivable (money owed a business)

Commission—pay based on a percentage of items sold rather than hours worked

Committed—will sacrifice to accomplish goals; is emotionally bound to the activity

Communication—oral, written, or visual messages between people

Company benefits—compensation aside from salary or wages

Company policy—guidelines followed by the company

Compassion—consideration or empathy for employees

Compatibility—the ability to work together in harmony

Compensation—the reward given employees for their work, usually in the form of wages, salary, benefits, and so on

Competition—businesses providing the same or similar products or services

Competitive markets—there are multiple bidders for the product being marketed

Competitor—a business providing the same product, service, or sales as another

Computer locater service—a computer search program. For example, all car stocks were listed on the dealer computer network and dealers could find desired units by conducting a search

Confrontation—head-to-head disagreement or quarrel

Congested—heavy; hindering free movement

Conservative—on the low side as compared to overestimating sales returns

Consideration—items to be carefully studied

Considered—thought about

Consume—use up, such as using money to buy a pizza that you eat

Consuming—using up; for example, you eat, or consume, a candy bar

Consummating—completing

Contact persons—people who know a business well enough to provide useful information

Contingency—preparation for a possible event

Continuous life—ongoing without an ending date

Convert—change

Cooperative—a business organization owned and operated by and for the benefit of its members

Copyright—exclusive right to publish and sell work for 28 years

Corporation—an entity owned by shareholders that conducts business as an individual

Cost—the amount paid for a resource

Cost benefit—a comparison of cost and return

Cost of goods sold—total wage and material resources for goods sold

Covenant—agreement, in this case not to compete

Creative person—one who has new ideas

Credit—obtaining goods or services on the basis of deferred payment

—an addition to a revenue, net worth, or liability account or a deduction from an expense or asset account (entry in the right hand column of the journal)

Credit agreement—written document setting forth credit terms

Credit application—a form used to gather credit decision information

Credit bureau—an organization that supplies credit information to business people

Credit card—bank cards or plastic money issued by many companies

Credit customers—customers who buy now and pay later

Credit extension—credit granted for a period of time

Credit limit—the dollar amount a supplier will extend on credit

Credit policy—the rules set for when or when not to grant credit

Credit rating—the evaluation of a person's ability to pay

Customer profile—description of your clientele to aid in decision making

Customer relations—plan for maintaining favor with customers

Data terminal—a computer terminal that interconnects to the central terminal with larger memory

Debit—an addition to an expense or asset account or a deduction from a revenue net worth, or liability account (entry in the left column of the journal)

Debt service—money needed to pay debts on time

Debts—money owed to others; liabilities

Deductible—the dollar amount specified in a policy which will be assumed by the insured in the event of a loss

Deflation—loss in money value due to the decreasing level of the economy

Delegation—having someone else be responsible

Delivery expense—cost of delivering

Demand—willingness to buy

Demographic—having to do with characteristics of a population; statistics which include size, growth, density, and distribution

Demonstrations—live animals fed in the store

Depreciation—an allowance made for the decrease in value of an asset because of wear or age

Differentiate—cause to be different than competition

Direct labor—the labor used in production (does not include administrative labor)

Direct mail advertising—mailed to prospective customers

Direct materials—the materials used in production (does not include materials for building maintenance, and so on.)

Direct writer—sells insurance for one major company

Disclosure—something revealed or made known

Discount—a reduction from the full cost of a price, usually given as a benefit to employees, or for multiple sales

Disk drive—computer input device

Distraction—something that diverts attention or concentration

Distrust—suspicion

Diversification—adding other products or services to improve sales volume, perhaps to improve seasonal volume

Dividends—earnings distributed to corporate stockholders

Doers—individuals who get things done

Domain—working area of responsibility
—affective psychological area dealing with feelings or emotions
—cognitive psychological area of the intellect
—psychomotor psychological area dealing with physical manipulative skills

Dot matrix printer—computer printer that uses pins to form letters, hence the term dot matrix

Double entry bookkeeping—record system that has both debit and credit entries that balance

Down payment—the capital invested at the time of a purchase, with the balance of the purchase to be paid in installments

Dreamer—an individual who has ideas but fails to carry them out

Earnings—money above expenses

Easement—a permanent right one has on land of another, for example telephone or power lines

Economic concept—supply and demand are two economic concepts

Economic downturn—tough times financially

Economic opportunity cost—comparison of returns against an alternative use of your resources

Economic reason—financial reason

Economic risk—the chance for financial loss

Economic system—financial system; we call ours capitalism

Economics—the study of efficient allocation of resources

Economist—one who studies economics

Elastic demand—sales that are greatly influenced by price

Emotional motivation—the desire to act which is caused by feelings or sensitivities

Emphatic—forceful

Employee discounts—lower prices allowed to employees by the owner of a business

Employee incentives—employer inducements to encourage employee productivity

Endeavor—task or undertaking

Entity—something that exists independently

Entrepreneur—a person who organizes, manages, and takes the risk in business

Enterprise—profit-seeking activity, a business

Equal Pay Act of 1986—a federal law requiring equal pay to women who perform the same tasks as men

Equity—net worth or that portion of a business that represents ownership above investment

Established—already in operation

Estate—property owned by a person

Estimate—calculation of cost for a customer

Ethics—standards of conduct that regulate one's actions

Excise tax—a tax on manufacture, sale, or consumption of a commodity such as tobacco, beer, and wine

Executive officers—operating or managing officers

Executive Order 11246—a federal law that requires no discrimination in employment practices on basis of race, sex, color, or religion

Expected—anticipated or projected

Expenses—costs of doing business

Expertise—high level of skill or knowledge

Extermination company—a private business providing pest control services

Event—the recognition of a change in resource cost within an entity

Facilities—the physical property such as land, buildings, equipment, and utilities

Factory overhead—costs that are independent of production level

Family living costs—money needed to pay the family's bills

Family corporation—a business owned by family members

Feed efficiency—pounds of grain needed to produce a pound of weight gain

Field representative—representative that works away from the home office; may have a specific territory to supervise

Finance charges—costs due to extending credit

Financial accounting—preparation of reports of financial history for people outside the business

Financial capability—ability to pay the stockholder or lender

Financial restrictions—limited use of money

Financial statements—a balance sheet; a financial picture at one point in time of a business required by a lender

Financial transaction—an exchange of money for goods or services between entities

Finished goods—inventory assets ready for sale

Fiscal year—a business year other than a calendar basis

Fixed assets—items that are not readily marketed, such as buildings

Fixed costs—costs that occur regardless of sales or production activities, for example, minimum electrical bills

Fixed interest rate—an interest rate that cannot change during the life of the loan

Flexible hours—variation of hours worked

Floor samples—items displayed so customers can examine them

Forced sale—sale caused by something out of the owner's control, for example, death of a partner

Forecasting—to estimate or calculate in advance; to predict future business activity by analyzing potential markets

Fortune 500—a listing of the top 500 businesses in the United States

Founder—the person or persons who start a business

Fringe benefit—compensation other than money, such as use of a company car, medical insurance, or other similar perquisites

Funded—financed or supported by money

General ledger—a loose-leafed or bound book with financial pages for accounting

Goals—objectives we try to achieve

Goods in process—inventory assets not now completed

Goodwill—the reputation of the business and the attitude of the public toward the business that cause them to do business with the previous or new owners

Grain shrinkage—reduced amount due to handling losses and moisture content

Graphics—the ability of the computer to display or print tables and pictures

Grievance—complaint

Group insurance plans—insurance provided by the employer in whole or in part

Happenstance—chance event

Hard disk—large supplementary computer memory

Hardware—the computer, printer, monitor, modems, disk drives, hard disks, joysticks, and similar items that operate the computer

Heirs—people who will inherit property

High cost producers—producers whose costs are higher than those of their competitors

Human motivation—the drive for behavior which is caused by human characteristics such as the quest for power, the need for a feeling of belonging, or the desire for profit

Human relations—the ability to work with others comfortably without friction

Hypothetical—assumed or made up

Image—how people view the business

Imagination—creative thinking

Immediate goal—objective to accomplish in less than a year

Incentive—something which causes action such as the expectation of a reward or the fear of punishment

Independent agent—one who sells for a number of companies

Indirect labor—any labor which is not used directly for production

Indirect materials—any materials not used directly for production

Inelastic demand—sales that are minimally affected by price

Inflation—increase in money value due to the increasing level of the economy

Inheritance tax—tax on assets passed from one generation to another

Input—data typed into the computer or loaded from a program disk

Inside information—things known to only a few people

Installment plan—purchase agreement for time payments

Installment purchase—a purchase which is not paid for outright; the installments are specified amounts paid at specified times

Insurable—a risk that an insurance company is willing to accept for premium payment

Insurance agent—sales representative who sells insurance and serves customers needs

Insurance claim—demand by an individual to recover a loss under an insurance policy

Integrity—adherence to a code of values

Interest free period—a period of time before interest will be charged on a credit purchase

Interest payment—the cost of using money, the return for letting others use your money

Intermediate assets—assets that would require a longer time for conversion

Intermediate liabilities—money due over a period of a year or more

Interpersonal relationships—ability to work with and motivate other people

Inventory—merchandise on hand

Investing—putting money in the bank to collect interest or into a productive enterprise

Investment tax credit—a tax deduction given for the purchase of business-related assets

Investments—money or material used to produce a product or interest

Invoice—price charged the dealer by the supplier

IRAs—Individual Retirement Accounts. These are government approved investment plans that offer tax advantages to employees

Isolated—located on a low traffic road

Job description—listing of duties for a job holder

Journal—a book in which transactions and events are entered; the entry includes the date, the description of the transaction or event, and the monetary results

Judgment—decision or opinion given after consideration

Laser printer—newest computer printing technology

Leader pricing—a technique used in retailing; an item is sold at or below cost to attract customers

Leave of absence—extended time off granted to an employee for special circumstances, such as maternity leave, illness, and so on.

Pay and benefits may be reduced or suspended during this time

Legal advice—advice from a person trained in the law

Legal compliance—following federal, state, and local laws

Lenders claims—the claims lenders have against the assets of a business

Lending committee—the committee in a lending institution that recommends that loans be granted or denied

Lessee—person or business using a rented asset

Lessor—owner of a rented asset

Liabilities—amounts that are owed

Liability insurance—any form of insurance coverage that protects against claims

License fee—fee paid for the privilege of operating an enterprise such as a taxi service or barber shop; some may be called permits

Liquidation—the process of selling off assets

List—put up for sale

Location—the physical placement of a business is usually one of the most crucial factors in success

Logical steps—well thought out, sensible plans to follow

Long-term goals—the goals that excite the individual; may take a period of years to complete

Long-term liabilities—debts due over a period of years

Loss frequency—how often losses occur

Loss severity—the extent of damage

Low cost producers—those producers whose costs are lower than their competitors' costs

Lower margins—adjusting price down so there is less profit

Mail survey—questionnaires sent by mail

Major—of national repute

Manageable units—dividing into smaller pieces to be completed

Management—the control of business decisions that have a large effect on success

Markdown—a reduction from the original price of an item, usually as a sales incentive

Market (marketing)—selling one's product or services

Market economy—an economy that anyone can enter freely

Market niche—your piece of the pie or a defined segment of a market

Marketing plan—plan for sales

Market pricing—setting the price at which items will sell

Market research—gathering and interpreting data about marketing

Market value—the price at which goods will change hands

Markup—an increase in the original price of an item to provide a profit at a retail sales level

Material resource—physical inputs to production

Maxim—a saying expressing an inherent truth

Maze—confusion network

Media—the means of mass communication, including television, newspaper, and radio

Memo—paper form used for brief notes between offices

Memory—computer's capacity to store information

Meticulous—accurate and detailed

Microfilm—small microphotographic negatives that reduce record bulk

Misrepresentation—statement as fact of an untruth

Modification—change

Monthly statement—a monthly record of purchase charges and payments

Motivator—something that causes one to take action

Multiple pricing—two for $1.98, and so on

Negligence—omission or neglect of reasonable caution

Net worth—owner's equity

Network—two or more computers interconnected

Nonbusiness entity—nonprofit group such as a church

Objective—something that serves as a goal or is the object of a course of action

Obsession—a preoccupation with a fixed idea

Obsolete equipment—equipment that is no longer needed or useful

Odd-ending pricing—a pricing method, for example, $1.98, not $2.00

Office politics—the behavior of employees in offices which is structured to create good relationships among peers and superiors; often intended as a derogatory description when such behavior is seen as phony

Open account—allows purchases on credit

Open house—special promotion to attract visitors to your business

Operating loans—open credit to permit day to day purchases of supplies and merchandise

Operating system—each computer has an operating system often known as DOS (Disk Operating System)

Options—a legal agreement that sets the right of a person to purchase a property at a later time

OSHA—Occupational Safety and Health Agency; federal agency that regulates safety in the workplace

Outdoor advertising—signs and billboards

Output—the results of computer use, may be printed on the screen or on paper by the printer

Outside changes—changes in the market due to unrelated events

Outside salesman—a person who does sales work away from the business location

Overhead—a commonly used term for fixed expenses; includes rent, utilities, taxes, and excludes direct labor and materials

Owner equity—business net worth

Owner's claims—what is left of assets after all claims against the business are paid

Package price—a single price covering a number of items

Paid-in capital—money invested by owners

Paper trail—documents used to prove balances

Partnership—two or more people operating as an entity

Past performance—how well an individual or business entity has met responsibilities; best indicator of future performance

Patent—right granted an inventor to exclusively make and sell his invention

Patronage dividends—rebate or bonus based on volume of purchases

Patrons—customers

Percentile—the number out of 100 people being compared

Peril—a cause for a possible loss, such as wind, fire, or hail

Period reports—profit and loss statement reports

Peripherals—devices that connect to the computer, such as printers, modems, and so on

Permanent records—records kept for several years

Perpetuation—to keep the business going

Perseverance or persistence—ability to keep going in spite of difficulties

Personal entity—any individual

Personal value—what a resource is worth to you without considering market value

Personnel review—job performance review

Pest control—control of insect, rodents, and other pests

Petty cash—small cash fund for minor purchases

Philosophy—what you believe

Physical motivation—the desire to act which is caused by physical needs such as hunger or discomfort

Planning commission—local group that oversees orderly development

Posting—recording sales tickets in the business's ledgers

Potential—can become possible

Precise—exact or accurate

Premium—the amount of money paid for an insurance policy

Preplanning—preparation and planning that goes into the enterprise before the business opens its doors

Present market value—what an item would sell for on the open market

Preservation—maintained rather than destroyed

Price—the cost at which something can be obtained

Price concession—lowering the price

Price leader—a business which emphasizes higher pricing due to professed quality

Price sensitivity—product or customer easily affected by price

Pricing strategy—carefully planned price of an item

Prime rate—interest rate banks charge their best commercial customers.

Priority—precedence, especially by order of importance or urgency

Printer—hardware device that types or prints computer output

Probable—likely

Problem solving—ability to analyze a situation and provide workable solutions

Procedure—a way of doing anything

Production volume—the number of items produced during a specified time

Procrastinate—to put off doing something until a future time; to postpone or delay needlessly

Product liability insurance—protection against claims arising from use, handling, or consumption of the product

Profit—the return received on a business undertaking after all operating expenses have been met

Property tax—a tax based on the value of real estate or other property

Prospective—not yet participating

Prospers—earns enough money to be profitable

Protected—kept safe

Prudent—wise or cautious

Public relations—fostering goodwill

Publicity—information that concerns a business which is passed along through various communications in order to attract public attention

Pure risk—an insurable loss

Purveyors of Doom and Gloom—those who are always negative

Rank—listing from least to most important

Rate of turnover—how frequently a given item sells

Ratio—a way of comparing things

Raw materials—inventory assets used in making goods to be sold

Real property—land and buildings

Realistic—practical, sees the most likely outcome and does not expect too much

Reasonable price—agreed to by both parties

Recapitulate—restate or tell again

Recipes—formulas or budgets; here used to describe how resources you now have can be combined to yield the most of what you want

Records—written or electronic notes of a transaction

Regress—to lose progress or go backwards

Regulations—governmental rules governing business activities

Relocation—required move to a new community

Remitting—paying the due amount

Renewal option—the opportunity, usually specified in a contract, to renew an agreement

Replacement—to provide an equal or better item for the insured item

Replacement cost—the cost of replacing damaged property

Reputation—how others perceive a person or business

Residual—left over or remaining

Resource—what it takes to operate a business, for example, mind, money, and muscle

Resume—a compilation of your experience, training, and personal references

Retrieve—to find again, such as from a file

Risk—chance for profit, loss, or mischance

Risk avoidance—how to avoid risk

Risk management—planning for risk reduction or insurance

Risk reduction—how to reduce risk of a loss

Risk retention—carrying the risk instead of buying insurance

Risk transfer—insurance purchase made to share risk with an insurance company

Running balance—current or up-to-date status of an account

Runts—unthrifty pigs that are underweight for their ages

Sales quota—an anticipated level of sales which serves as a goal

Sales slip—details of a transaction

Sales supervisor—a person co-ordinating the work of sales people

Sales volume—the number of sales made during a specific time

Sales zones—that area from which a business hopes to attract customers

Sanitation products—livestock supplies

Scarce or scarcity—not readily available; money supplies illustrate the concept that most of us do not have all the money we want so we must allocate it for what we want most

Scheduling—planning time for specific activities

Seasonal items—items needed only part of the year

Secure—safe; lockable to prevent unauthorized entry

Self-discipline—doing the required activities on time without others telling you

Seniority—longest time on the job

Service club—an organization of business people who work for fellowship and community improvement

Setback—the distance the building need be from the street or property line

Severance pay—pay settlement when an employee has been terminated

Short-term goal—a goal to be accomplished in 1–5 years

Silent partners—nonoperating partners

Single entry bookkeeping—simplest accounting system

Skill—an ability to do something

Small Business Administration—a U.S. government agency whose purpose is to promote small business development

SOEP—supervised occupational experience program; the application step in vocational education

Software—programs usually on diskettes that perform the desired tasks on computers

Solvency—good financial position; if assets were sold, debts could be paid

Speculative risk—a risk taken in which an individual can profit or lose on a business venture; not insurable

Spin-off—a new business started from an established business

Spouse—husband or wife

Start-up business—early stages of a business venture

Start-up capital—money needed to start a business

Statutes of limitation—permissible time frame for legal action

Stock appreciation—stock value increase

Stock company—a corporation; ownership by stock shares

Stock shares—certificates of ownership in a corporation

Stockholders—owners of a corporation

Strategies—plans for action

Structure—building
—how a business is organized

Subcontracting—having a second company do a portion of a contracted job

Suppliers—businesses who provide products to the business for use or resale

Supply shifters—changes in cost of production, availability of raw materials are examples

Support—help in operational problems as well as the repair and maintenance capabilities

Support group—people who are helpful and encouraging

System—the computer and peripherals

Task—an identifiable unit of a job holder's work that can be broken down in steps from a beginning to an end

Taxation—assessing payments from individuals to pay the cost of government

Technology—changes such as computer operated machinery

Ten Most Wanted—an organized method of arranging our tasks, selecting the 10 with highest priority

Tenacity—persistently maintaining effort over extended periods of time

Termination—bring to an end

Time management—how one uses the 24 hours in a day

Top 10%—many companies recognize the performance of their outstanding sales personnel through incentive plans—vacation trips for salesperson and spouse, cash bonus, gifts, and so on

Trade shows—public shows displaying new or recent items of a given kind

Trading area—the area surrounding a city or town that draws customers to the city or town

Traffic—activity in an area; two types are important—auto (people passing through the area) and foot (those shopping in the area)

Traffic patterns—how traffic moves in an area of a business location

Training program—a structured learning program to prepare new job holders

Transactions—individual sales or purchases

Types of credit—ways credit is granted

Ultimate—of gravest or greatest importance

Unexpected expenses—unanticipated costs

Unflagging—continuing, never ending

User fee—money paid by the user of a service

Utility—value other than money

Value leader—a business which emphasizes lower pricing or value buying for its customers

Variable cost—the cost of labor or material which increases with every item produced or every service rendered

Vendor—person selling a product or service

Venture—an undertaking such as starting a business

Venture capital—capital one has available to risk on new ventures; some people make a business of providing venture capital

Viable—existing as an independent unit

Ward—dependent being; one supported by others

Wealth—possession of money or material assets that could be sold for money

Widget—a generic term for an imaginary product

Word of mouth—in advertising, we are talking about referrals, in other words, one customer refers a prospective customer to your business

Worker's compensation—a system providing for payments to employees because of injury on the job

Work permit—provided by a school administrator for underage workers

Zoning restrictions—regulations for a certain area; for example a fence around a junkyard, and so on

INDEX

A

Accountant, 109, 113–15, 120
Accounting, 109–114, 121, 122, 126–38
    accrual, 121
    cash, 121
    computer, 120
    financial, 137
    practices, 110, 111
    rules, 111
    system, 122
Accounts
    new, 88
    payable, 160
    receivable, 161
Advertising, 22, 95
Appraisers, 113–15
Assets, 23, 62, 131–35
    current, 23
    intermediate, 24
    long-term, 24
Auditing, 137

B

Bad debt losses, 8
Bank account
    business, 122
    setting up, 122
Balance sheet, 23, 62, 128, 135, 157
Benefits, 275, 279
    flexible, 279
    general, 279
    managing, 279
Bookkeeping, 120, 127, 137
    double entry, 118, 120, 125–27
    single entry, 118–22
Break-even analysis, 220–21
Business
    buying, 52, 53, 56, 57
    entity, 125, 126, 131
    failures, 288
    finances, 23
    image, 55, 101
    incentives, 319

    location, 22, 25, 55, 190–95
    perpetuation, 317
    plan, 16, 19, 312
    policy, 243–45
    regulation, 194
    selling, 323–27
    types, 7
    venture, 7, 16, 27

C

Capital, 16, 236
    paid in, 132–33
Cash flow, 16, 24, 146, 160–62, 234–36
Claims, 131
Commitment, 6
Communications, 246, 248, 274
Compensation, 275
Competition, 8, 21, 70–76, 193
Company innovators, 7
Computers, 300–305
    data terminal, 300
    hardware, 300
    modem, 305
    printers, 304
    selection, 304
    software, 300, 301
Consumption, 64
Cooperatives, 7, 44
Copyrights and patents, 178
Corporation, 7, 44, 46, 47, 327
Costs, 204–215
    direct, 211
    fixed, 209–210
    fixtures, 207
    hidden, 206
    indirect, 211
    inventory, 207
    legal, 207
    overhead, 208
    start-up, 207
    unexpected, 208
    variable, 210
Cost of goods, 136

Credit, 165–73
    advantages, 168
    application, 171
    cards, 169
    charge accounts, 168
    collections, 171
    disadvantages, 168
    installment, 168, 236
    policy, 172
    regulations, 170, 237
    revolving, 169
    terms, 236
    types, 168, 236
Credits, 129, 130
Customer development, 88, 96, 258–63

D
Debits, 129–30
Debts, 62
Decisions, 85, 110, 144–51, 156–62
Delegation, 252
Demand, 72, 73, 75, 226
    elastic, 73
    inelastic, 73
Depreciation, 250
Developers, 7
Doers, 27
Dreamers, 27
Duplicators, 7

E
Earnings, 132–33
    retained, 132–33, 137
Easements, 178
Economic cycle, 8
Economics, 110
Economist, 113–15
Employee, 22
Employment
    application, 246
    job description, 275
    satisfaction, 280
Entities
    business, 125, 126
    non-business, 126
    personal, 126
Entrepreneurial traits, 5, 348, 349
Events, 127, 134, 194

Expansion 86, 309
    product, 310, 311
    service, 311
Expenses, 136, 208
    advertising, 208
    delivery, 208
    insurance, 208
    interest, 209
    maintenance, 209
    salaries and wages, 208
    supplies, 208
    taxes, 209
    utilities, 208

F
Facilities, 193, 197–200
    buy or rent, 200
Financial
    daily summaries, 121
    income statement, 23, 158
    monthly summaries, 121
    period reports, 161
    plan, 16, 23, 25
    trial balance, 135
Franchise, 47

G
Goals, 32–38
    business, 37
    daily, 38
    immediate, 33, 36, 37
    long-term, 34, 36, 37
    setting, 32, 33
    short-term, 29, 33, 37
    ten most wanted, 36
Grievances, 275

H
Human relations, 261, 267
Hiring procedures, 246, 274

I
Independence, 6
Independent agent, 293
Insurable risks, 290
Insurance, 283–95
    arbitration, 295
    coinsurance, 294

deductable, 294
health and life, 292
liability, 291
loss frequency, 295
loss severity, 295
property, 293
selecting companies, 295
Interviews, 277
Inventors, 7

J

Job
description, 275
evaluation, 278
pricing, 278
Journal, 129, 130

L

Leases, 177, 213
advantages, 214
agreements, 213
disadvantages, 214
Legal, 176–81
Liabilities, 24, 62, 131–35
current, 24
intermediate, 24
long-term, 24
Licenses, 178, 207

M

Management,8, 107
Market niche, 80–89
Market research, 82–87
interviews, 84
mail surveys, 84
telephone, 84
Market strategy, 97
Marketing plan, 22, 25
Microfilm, 122
Motivation, 3, 30, 31
emotional, 30
human, 30
physical, 30
Media
billboards, 100
business cards, 100
catalogs, 99
circulars, 99

demonstrations, 99
fliers, 99
magazines, 98
mail, 99
newspapers, 98
radio, 98
signs, 100
television, 98

N

Net worth, 62, 64, 131, 135

O

Optimism, 6
Options, 177
Ordinances, 178, 194
Organization, 5, 23, 254
Organizational levels, 254
Occupational Safety and Health Administration
(OSHA), 177

P

Partnership, 7, 42, 44, 46, 47, 326
agreement, 319
silent, 42
Persistence, 6, 341
Personal liability, 317
Personal value, 109
Personnel review, 275
Petty cash, 121, 250
Present market value, 107
Present personal value, 107
Prices, 218–28
changing, 227
competitive, 227
considerations, 218, 224
contracting, 228
differences, 228
establishing, 227
manufacturing, 228
market, 227
regulations, 226
sales, 228
sensitivity, 228
service, 228
strategy, 228
Priority, 35
Procrastinate, 34

Problem solving, 5, 239, 240
Production, 7
Profit, 133, 222
Promotion, 94–101, 275
Public relations, 101, 265, 267
    greeting, 267
    helping, 268
    listening, 268
    performing, 269
    serving, 269
    smiling, 267
Publicity, 100, 265
    charitable organizations, 101
    civic activities, 101
    community events, 100
    open houses, 100
    word-of-mouth, 101, 258

R
Recordkeeping, minimum, 248
Records, 104–115,118–22, 157, 158, 248
Record preservation, 121
Reports, 135–36, 157–62
    point in time, 135
    period, 136
Resources, 59–66, 109, 110, 131
    mind, 4, 5, 65
    money, 4, 5, 64
    muscle, 4, 5, 65
Responsibility delegation, 253–54
Retirement, 225
Revenues, 136
Risk
    avoidance, 289, 340
    insurable, 290
    management, 287, 288
    pure, 291
    reduction, 340
    retention, 289, 341
    taking, 288
    transfer, 290, 340

S
Sales, 7
    forecasting, 223
    quota, 224
    reports, 158
Scalpers, 7
Scheduling production, 232
Service, 7
Small Business Administration (SBA), 21
Sole proprietorship, 7, 43, 46, 47
Statute of limitations, 121
Subcontracting, 215
Supply, 74, 75, 226
    shifters, 75

T
Taxes, 184–87
    fairness, 185
    property, 172, 184
    sales, 177
    user fees, 184
    withholding, 176
Termination, 275
Time management, 34
Time off, 275
Training, 245, 275
Transaction, 64, 127, 134
Trial balance, 135

V
Vacation, 275
Venture capital, 125
Vision, 5

W
Work habits, 39
Work permits, 178
Worker's compensation, 177

Z
Zoning, 21, 178, 195